SURMOUNT
STYLE+COPY II 突破风格与复制 II（上）

香港建筑科学出版社
唐艺设计资讯集团有限公司 策划

香港建筑科学出版社 编

天津大学出版社
TIANJIN UNIVERSITY PRESS

PREFACE

序言

This is a chaotic and complicated information age, in which people are overwhelmed by large amount of mixed information from all times around the world. Now the distinct line between time and space becomes hazy; classic and avant-garde turn to be short-lived. People get tired and lost under the fast changing of aesthetics, value, life style, and etc. If residential design is a lifestyle customization which will fix someone's living way, its key words should be how to balance the relationship among society, human and family.

Nowadays, China and many developing countries are in the high-speed developing period, referring to not only economy, but also rapid changing in value and aesthetics for growth of knowledge and culture. How can residential design not only serve local people, but also satisfy international needs, not only reflect the features of times, but also back to people-orientation? This requires designers pay more attention to society and broaden their knowledge. Residence design has exceeded its own meaning and value; it not only has fundamental and completed functions, but also need to have more added values in order to reflect the significance of the various levels, meet the changing trends and get close to the modern people on the present and spiritual needs for future living space.

Surmount Style+Copy has selected elaborately more than 80 outstanding works in major cities around the world designed by certain famous architects to fully demonstrate the tendency of global residence design. Architecture itself is an embodiment of thinking, which will show different meanings by different readers. We only have an analysis in basic aspects of façade, structure, ecology and sustainability, more deeper discoveries are waiting you to find in the book.

HongKong Architecture Science Press
Editorial Board

这一个混沌而繁杂的信息化时代，人们被穿越古今中外、形形色色的资讯所淹没，时间和空间变得朦胧，经典与前卫只是昙花一现。审美观、价值观、生活方式等快速地变化着，人们变得疲惫而迷失。如果住宅设计是一种生活方式的订制，如何平衡社会、人、家三者之间的关系则是住宅设计的焦点问题。

中国和很多发展中国家都处于高速发展期，不仅仅是经济的增长，知识与文化的增长也给价值观、审美观带来更快速的变化。如何做到既要服务当地群众，又要满足国际需求，既反映时代特色，又回归以人为本的本源？这需要建筑师更加关注社会、具备更宽广的知识层面。住宅设计已经完全超越其本身的意义和价值，它不仅要具有基础和完备的功能，还需要具备更多的附加价值，体现各个层面错综复杂的意义，更要面对不断变化的趋势，符合现代居住者以及对未来居住空间的精神需求。

《突破风格与复制》特意从世界著名建筑事务所中遴选全球主要城市的 80 多个精彩案例，是为了更加多元、全面地展示世界住宅设计的趋势。建筑本身就是一种思想的展示，不同的读者可以赋予它不同的寓意，我们仅从立面、结构、生态、可持续发展等基本方面加以分析，更多的、更深层的隐喻期待读者去发掘、去解读。

香港建筑科学出版社
编委会

CONTENT
目录

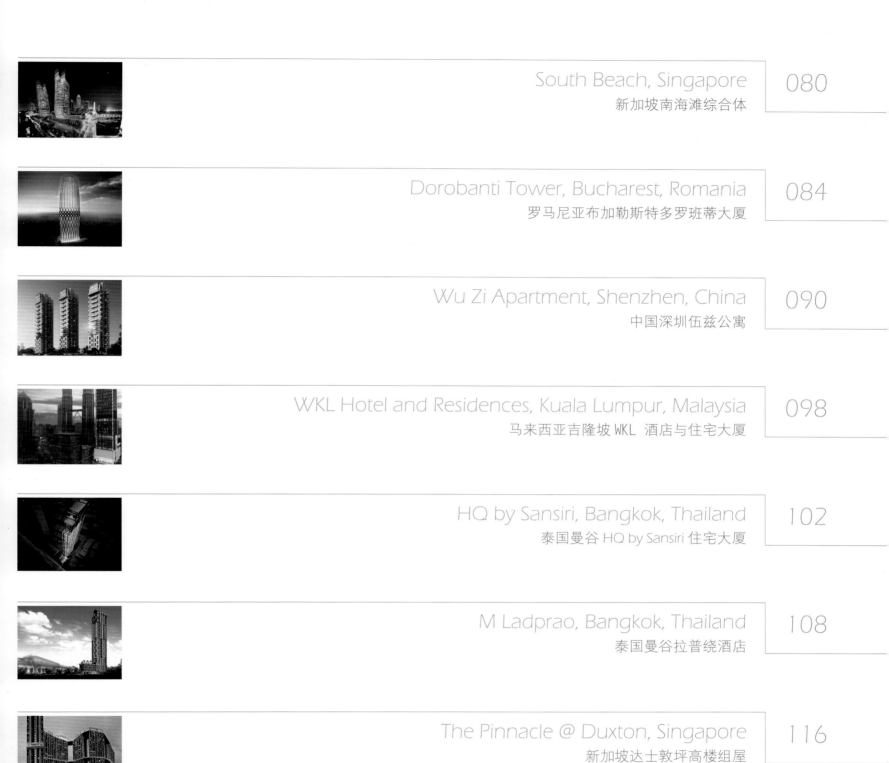

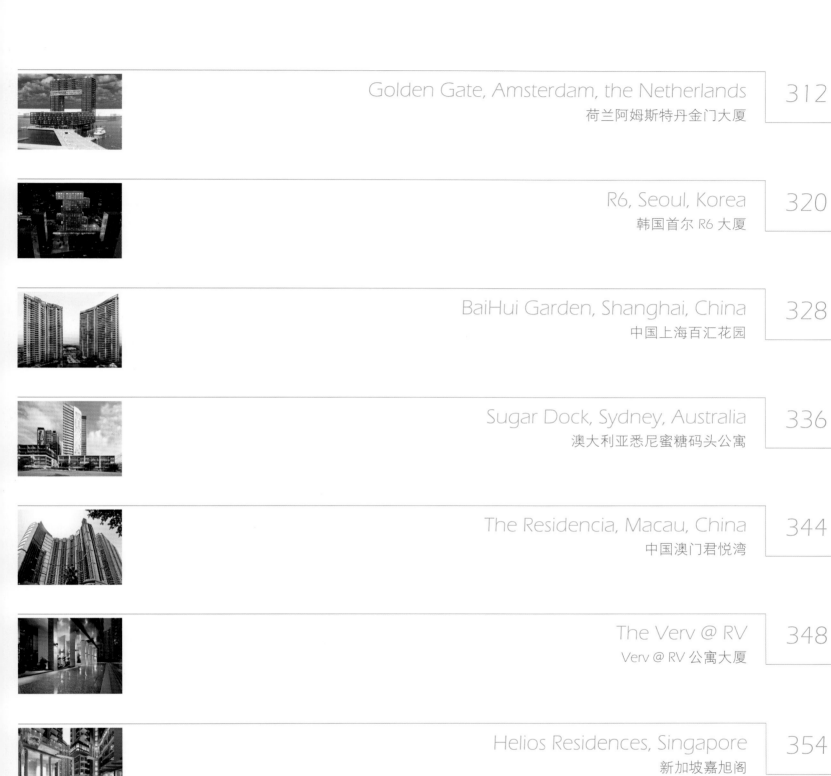

VIEW 视野

SHAPE 造型

FAÇADE 立面

GREENERY 绿色

Reflections at Keppel Bay, Singapore

新加坡吉宝湾映水苑

Architect: Studio Daniel Libeskind
Client: Keppel Land International
Location: Keppel Bay, Singapore
Site Area: 185, 806.08 m²
Floors: low layer 6–8, high layer 24–41
Photography: Keppel Bay Pte Ltd, VMW Obilia
Sketches and Plans: Studio Daniel Libeskind, Obilia

设计公司：丹尼尔·里伯斯金工作室
客户：吉宝置地国际有限公司
地点：新加坡吉宝湾
占地面积：185 806.08 平方米
层数：低层 6—8，高层 24—41
摄影：吉宝置地、VMW Obilia
草图和规划图：丹尼尔·里伯斯金工作室、Obilia

Prominently situated at the entrance to Singapore's historic Keppel Harbor, Reflections at Keppel Bay is a 185,806.08 m² residential development comprised of 6 high-rise towers ranging from 24 to 41 stories and 11 low-rise villa apartment blocks of 6–8 floors — a total of 1,129 units.

"吉宝湾映水苑"综合住宅项目耸立在新加坡吉宝湾沿岸，占地 185 806.08 平方米，包括六栋拔地而起的高达 24—41 层的高楼与 11 栋 6—8 层的低楼层公寓楼，共 1 129 套住宅单元。

The series of high-rise undulating towers is the focal point of this project. These sleek curving forms of alternating heights create graceful openings and gaps between the structures allowing all to have commanding views of the waterfront, Sentosa, the golf course and Mount Faber.

The design is composed of two distinct typologies of housing: the lower villa blocks along the waterfront and the high-rise towers which overlook them set just behind. The artful composition of ever shifting building orientations, along with the differing building typologies, creates an airy, light-filled grouping of short and tall structures.

These ever shifting forms create an experience where each level feels unique as it is not in alignment with either the floor above or below. No two residences alike are experienced next to one another or seen from the same perspective; the result of this design is a fundamental shift in living in a high-rise where individuality and difference is not sacrificed.

这一系列波浪起伏的高层大厦是这个项目的焦点。这些高低不一的光滑曲线形态营造出了优美的开口与缺口，这样的结构可以让人们一览海滨风情、圣淘沙岛、高尔夫球场和花柏山公园的美景。

设计是由两种类型的住宅组成，底层公寓置身高层公寓前方，沿水而建。方向各异的楼宇艺术性地组合在一起，随着建筑类型的变化，创造出一个空灵、闪亮的高层结构建筑。

这些不断变化的形式给人们这样一种体验：每一层都独具一格，与其他层不相一致。没有哪两座相似的建筑可以紧密相连或是可以从同一视角看到，这种设计的结果在高层居住中是一次基础性的转变，在这次转变中仍保留着个性与差异性。

FEATURE 特点分析

SHAPE

The sleek curving forms of alternating heights create graceful openings and gaps.

造型

不同高度的各部分建筑体因光滑的曲线造型而间隔，在高密度空间中创造出优雅的开口和间隙。

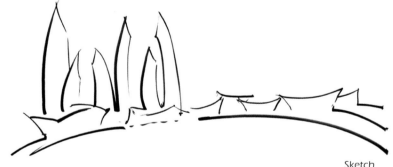

Sketch
手绘草图

Elevation
立面图

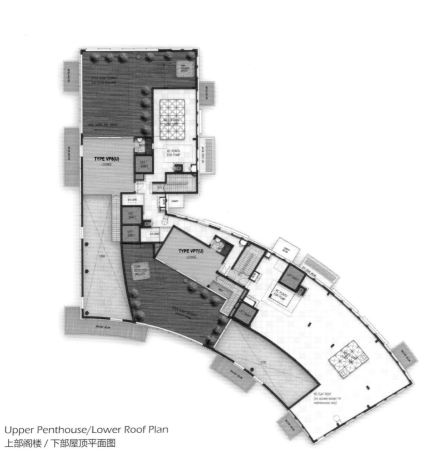

Upper Penthouse/Lower Roof Plan
上部阁楼／下部屋顶平面图

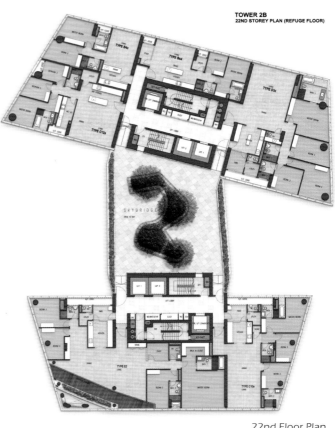

TOWER 2B
22ND STOREY PLAN (REFUGE FLOOR)

22nd Floor Plan
22 层平面图

The buildings are surrounded by an open architecture framework, continue the entire building form on the vision, and then the gardens on the top of them make a graceful ending. Pedestrian overpass connects between different floors of the building. Aluminium façade sparkling under the sun reflects changing sunshine and ocean scenes.

The project was completed in December 2011 and was the recipient of Singapore's Building and Construction Authority (BCA) Green Mark Gold Award and Universal Design Award (Silver). Additionally, Reflections at Kappel Bay has achieved top scores for the internationally-acclaimed CONQUAS® (Construction Quality Assessment System) by the Building and Construction Authority of Singapore.

高楼呈现出一个开放式的结构框架，在视觉上延续了整幢建筑的形式，屋顶花园形成了一个优雅的结尾。人行天桥在建筑的不同楼层相连，铝制立面在阳光照射下闪闪发亮，反射着不断变换着的日照与海洋景象。

该项目已于 2011 年 12 月竣工，并获得新加坡建设局颁发的绿色建筑标志金奖和通用设计银奖。此外，该项目取得由新加坡建设局授予的国际建筑质量评估体系 CONQUAS® 的高标准评价。

GREENERY 绿色

SHAPE 造型

FAÇADE 立面

ROOFTOP GARDEN 屋顶花园

The Interlace, Singapore

新加坡翠城新景

Lead Designer: OMA
Project Architect: RSP Architects Planners & Engineers (Pte) Ltd
Location: Singapore
Site Area: 81,000 m²
Gross Floor Area: 169,600 m²
Floors: 24
Images: CapitaLand
Drawings: OMA/ RSP
Realization: 2014

主设计公司：大都会建筑事务所
建筑公司：雅思柏设计事务所
地点：新加坡
占地面积：81 000 平方米
建筑面积：169 600 平方米
层数：24
图片：凯德集团
图纸：OMA/ 雅思柏设计事务所
完成时间：2014 年

The Interlace is one of the largest and most ambitious residential developments in Singapore. It sits on an elevated 81,000 m², 99-year leasehold site bounded by Alexandra Road and Ayer Rajah Expressway. The site completes a 9,000 m-long green belt that stretches between the Kent Ridge, Telok Blangah Hill and Mount Faber parks. With about 170,000 m² of gross floor area, the development will provide 1,040 apartment units of varying sizes with extensive outdoor spaces and landscaping.

翠城新景是新加坡目前最大型、最具挑战性的住宅发展项目之一。项目坐落于亚历山大路和亚逸拉惹高速公路交界处的高台地段，占地 81 000 平方米，租赁期为 99 年。该地块与肯特岗公园，直落布兰雅山公园和花柏山公园连接形成了一条延伸 9 000 米的绿化带。近 170 000 平方米的建筑面积可提供 1 040 个不同尺寸的住宅单位，且每一套都享有宽敞的室外空间及优美景观。

FEATURE 特点分析

SHAPE

The Interlace breaks away from the standard typology of residential developments in Singapore which comprises a cluster of isolated, vertical towers. Instead, its design explores a dramatically different approach to tropical living with an expansive and interconnected network of communal spaces with the natural environment.

Thirty-one apartment blocks, each six stories tall, are stacked in a hexagonal arrangement to form eight large-scale courtyards. The interlocking blocks resemble a "vertical village" with cascading sky gardens and both private and public roof terraces. Extensive residential amenities and facilities are interwoven into the lush vegetation and offer opportunities for social interaction, leisure, and recreation within the green terrain.

造型

翠城新景打破了新加坡住宅发展一贯的孤立、塔楼群的模式。反之，它的设计采用了一种别具一格的理念并体现出热带居住的方式，将自然环境与宽敞且丰富的公共空间格局相结合。

31 座六层高的公寓楼体按六边形形状摆放，形成了 8 个大尺度的通透庭院。交错的楼体好似一个"垂直村落"，重叠的空中花园以及公共和私用的露台延长了庭院的景观。大量的社区设施交织在茂盛的园林环境之中，为人们在绿色环境中提供了社交活动，休闲及娱乐的场所。

Site Plan
总平面图

N

0 5 10   20   30 Metres
Scale 1:600

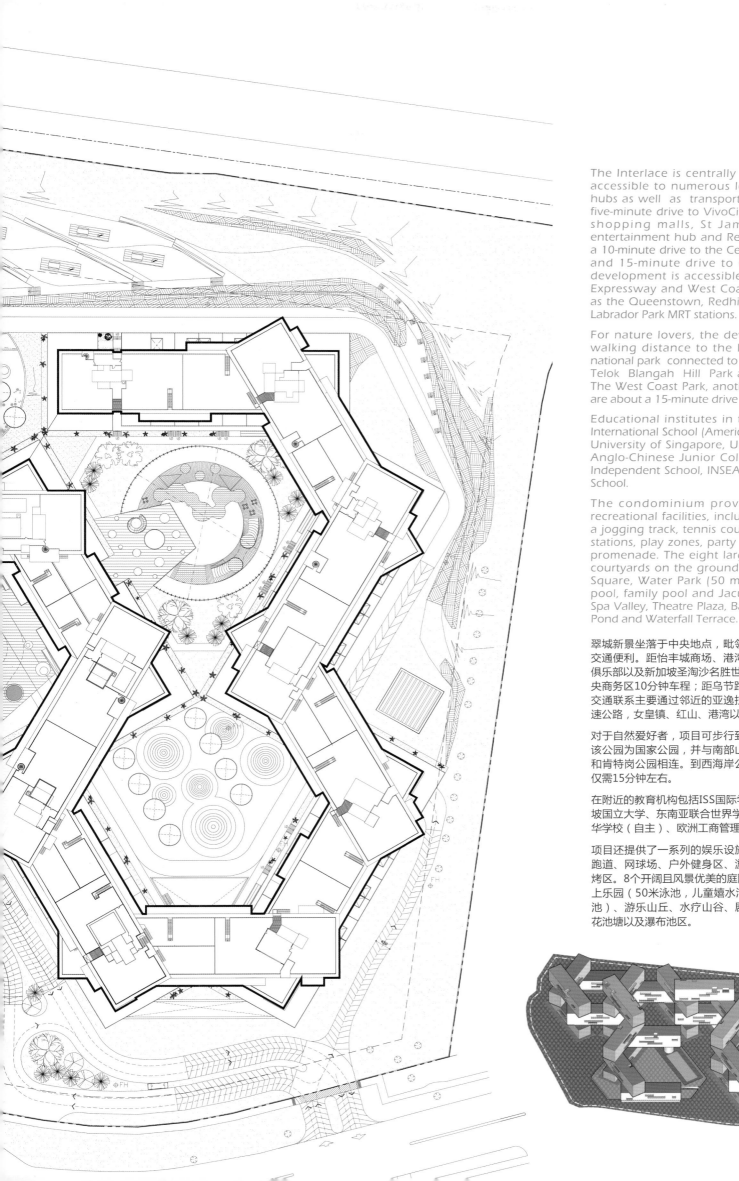

The Interlace is centrally located and readily accessible to numerous leisure and business hubs as well as transport connections. It is a five-minute drive to VivoCity and HarbourFront shopping malls, St James Power Station entertainment hub and Resorts World Sentosa; a 10-minute drive to the Central Business District and 15-minute drive to Orchard Road. The development is accessible via the Ayer Rajah Expressway and West Coast Highway, as well as the Queenstown, Redhill, HarbourFront and Labrador Park MRT stations.

For nature lovers, the development is within walking distance to the HortPark which is a national park connected to the Southern Ridges, Telok Blangah Hill Park and Kent Ridge Park. The West Coast Park, another recreational area are about a 15-minute drive away.

Educational institutes in the area include ISS International School (American College), National University of Singapore, United World College, Anglo-Chinese Junior College, Anglo-Chinese Independent School, INSEAD and Crescent Girls' School.

The condominium provides a full suite of recreational facilities, including the clubhouse, a jogging track, tennis courts, outdoor exercise stations, play zones, party pavilions and a BBQ promenade. The eight large-scale, landscaped courtyards on the grounds comprise a Central Square, Water Park (50 m lap pool, children's pool, family pool and Jacuzzi pool), Play Hills, Spa Valley, Theatre Plaza, Bamboo Garden, Lotus Pond and Waterfall Terrace.

翠城新景坐落于中央地点,毗邻多家休闲和商业中心,交通便利。距怡丰城商场、港湾商场、圣詹姆士发电站俱乐部以及新加坡圣淘沙名胜世界只需5分钟车程;距中央商务区10分钟车程;距乌节路15分钟车程。此外项目交通联系主要通过邻近的亚逸拉惹高速公路和西海岸高速公路,女皇镇、红山、港湾以及拉柏多公园地铁站。

对于自然爱好者,项目可步行到新加坡园艺园林公园,该公园为国家公园,并与南部山脊、直落布兰雅山公园和肯特岗公园相连。到西海岸公园和其他娱乐区域驾车仅需15分钟左右。

在附近的教育机构包括ISS国际学校(美国大学)、新加坡国立大学、东南亚联合世界学院、英华初级学院、英华学校(自主)、欧洲工商管理学院和克信女中。

项目还提供了一系列的娱乐设施,包括俱乐部会所、慢跑道、网球场、户外健身区、游乐区、派对亭台以及烧烤区。8个开阔且风景优美的庭院里建立了中央庭院、水上乐园(50米泳池,儿童嬉水池,家庭泳池以及按摩浴池)、游乐山丘、水疗山谷、剧院广场、青竹园林、莲花池塘以及瀑布池区。

Bird's Eye View
鸟瞰图

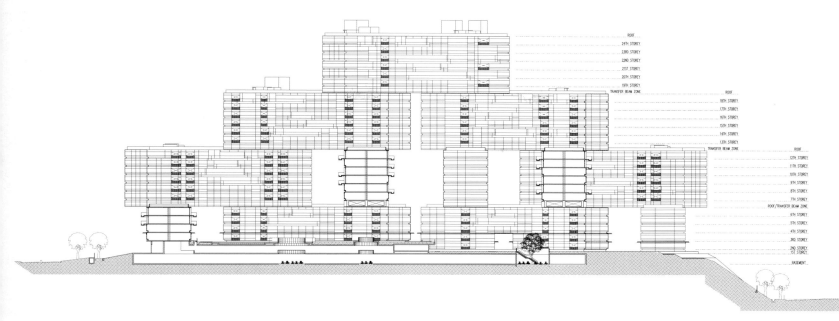

Section A-A
剖面图 A-A

South Elevation
南立面图

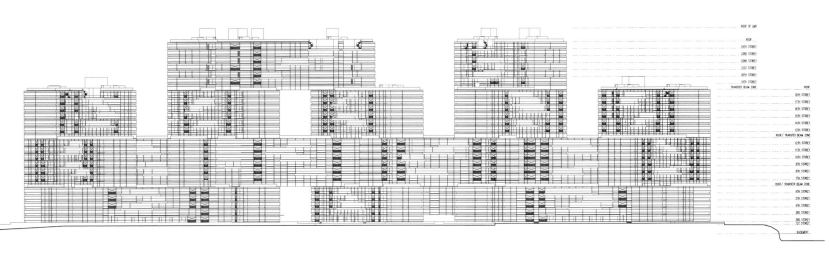

North Elevation
北立面图

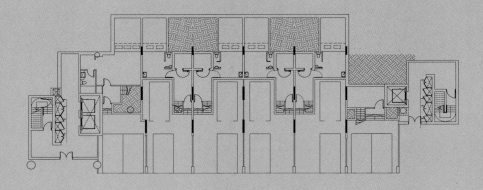

Block 1 & 2 - Basement
1 & 2 号楼地下室平面图

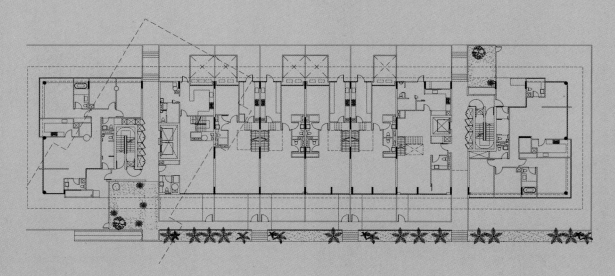

Block 1 & 2 - 2nd Storey
1 & 2 号楼 2 层平面图

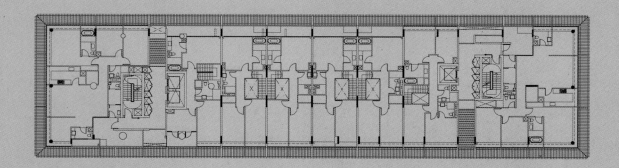

Block 1 & 2 - 3rd Storey
1 & 2 号楼 3 层平面图

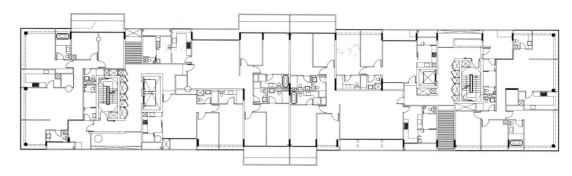

Block 1 & 2 - 4th Storey
1 & 2 号楼 4 层平面图

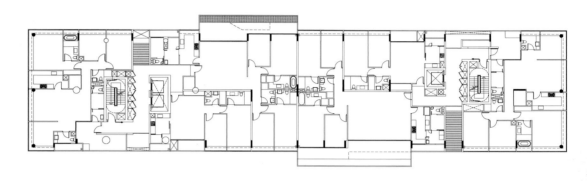

Block 1 & 2 - 5th Storey
1 & 2 号楼 5 层平面图

Block 1 & 2 - 6th Storey
1 & 2 号楼 6 层平面图

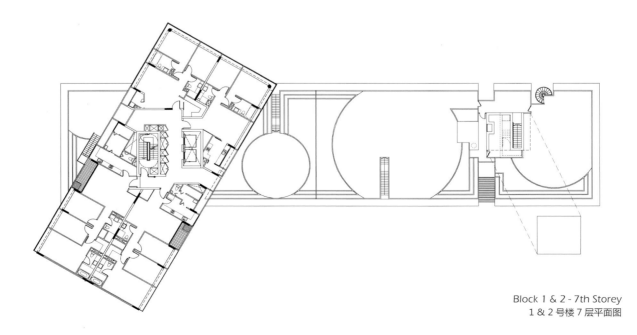

Block 1 & 2 - 7th Storey
1 & 2 号楼 7 层平面图

FAÇADE 立面

SHAPE 造型

STRUCTURE 结构

MATERIAL 材料

Baku Flame Towers

巴库火焰大厦

Architect: HOK	设计公司：HOK
Client: Azinko Development MMC	客户：Azinko Development MMC
Location: Baku, Azerbaijan	地点：阿塞拜疆巴库市
Floors: 33, 30, 28	层数：33、30、28
Height: 190 m, 160 m, 140 m	高度：190 米、160 米、140 米
Realization: 2012	完成时间：2012 年

Known as the "region of eternal fires", Baku's long history of fire worshiping provided the inspiration for the development's iconic design.

该区以"永恒的火焰地区"而闻名，巴库对火焰的崇拜的悠久历史为项目设计提供了灵感源泉。

SHAPE

The project consists of three flame-shaped towers, each with a different function, set in a triangular shape. Standing at 140 m-high, Baku Flame Towers will be seen from most vantage points within the surrounding area and is already a prominent feature of the Baku skyline.

造型

项目由3个火焰造型的塔楼组成，它们分别具有不同的功能，构成了一个三角形。火焰塔高达140米，在该区内最有利的地点都可以仰视到它们，构成了巴库天际线上一个独有的亮点。

HOK undertakes the master planning, concept and schematic design for the site and towers, and construction work is already well underway. The residential tower sits to the south, accommodating some 130 residential apartments over 39 floors, and is the tallest of the three towers. It will house luxury apartments and boast stunning views across the surrounding area, while the hotel is sited on the northern corner of the site, consisting of 250 rooms and 61 serviced apartments over 33 floors. The office tower is set on the western side of the complex, providing a net 33,114 m² of grade A flexible commercial office space.

HOK are currently working on the retail podium which will act as the anchor for the project, providing all of the leisure and retail facilities that will service the three towers and visitors to the development.

HOK 承担了项目场地和楼身的总体规划、概念和原理设计，建设工作已经起步。住宅大厦坐落在南部，高度超过39层，容纳约130所住宅公寓，是这三个塔楼中最高的建筑，坐拥豪华公寓和绝妙的景观。而酒店位于北部，高度超过33层，由250间客房和61所酒店式公寓组成。办公大楼位于建筑的西侧，提供了净面积33 114平方米的A等级灵活商业办公空间。

HOK 目前正在从事的零售区裙楼设计，将作为项目的一个支柱，提供所有休闲和零售设施，为三座大厦和游客提供了良好的服务。

MATERIAL 材料
FAÇADE 立面
SHAPE 造型
ROOF GARDEN 屋顶花园

Velo Towers, Seoul, Korea
韩国首尔维洛大厦

Architect: Asymptote Architecture   设计公司：渐近线建筑事务所
Location: Seoul, Korea   地点： 韩国首尔市

The Velo Towers are composed of a dynamic arrangement of two stacked and rotated volumes that are a formal and programmatic counterpoint to the conventional extrusion of massing that exemplifies the supertall as a building type. The Velo project proposes an alternative architectural and urbanistic response to the repetitive and monolithic austerity of conventional tower design. The Towers' 8 distinct residential components are rotated and positioned within a carefully choreographed massing arrangement, calibrating the orientation and views of each residential volume and taking full advantage of the Towers' position adjacent to the Yongsan Park overlooking the Han River in the distance.

维洛大厦由两座堆叠、旋转的体量以动感的排列形式组成，且与传统摩天大楼的形象形成正式又标志性的对比。本案突破了传统大楼设计重复又整体的内缩型特征，取而代之的是建筑化和都市化的表现。建筑有 8 个独立的住宅部分，以旋转的形式精心布局，保证每座住宅体量的朝向和视野，并充分利用了项目毗邻可俯视远处的汉江的龙山公园的区位优势。

With a collection of roof gardens, shared amenities and internal circulation around light-filled open atrium spaces, the vertically distributed massing elements create unique 6-to 8-storey residential communities on the skyline. The towers are joined by two bridge structures that house shared public amenities, and act as neighborhood scale "connectors" for the towers' residents. The building's raised plinth hovers above the communal landscape surrounding the base of the Towers while the skybridge floats 30 storeys above, housing fitness and recreation centers, lounges, pools, spas and cafes along with a sky garden providing spectacular views over the entire Yongsan site.

建筑集合了屋顶花园、公共设施和围绕着光线充足的开放式中庭空间的内循环系统，垂直分布的各个体块元素创造出了天际线中独特的6至8层的住宅区。两座体量以两座桥梁结构的部件相互连接，它们也作为公共设施和楼内居民的大型"联系纽带"。建筑高抬的底座漂浮在围绕着塔楼底部的公共景观之上，而连接的天桥结构则位于其上面的30层处，内含健身房、娱乐中心、休息处、游泳池、水疗中心、咖啡厅和能欣赏整个龙山区壮阔景色的空中花园。

Concept Sketch
概念草图

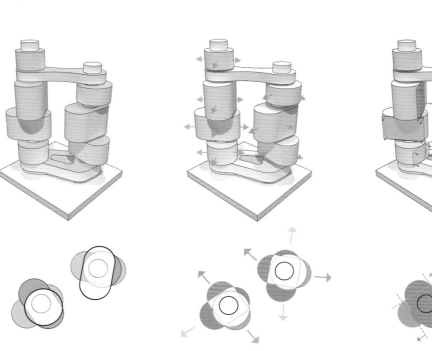

Structure Analysis
结构分析图

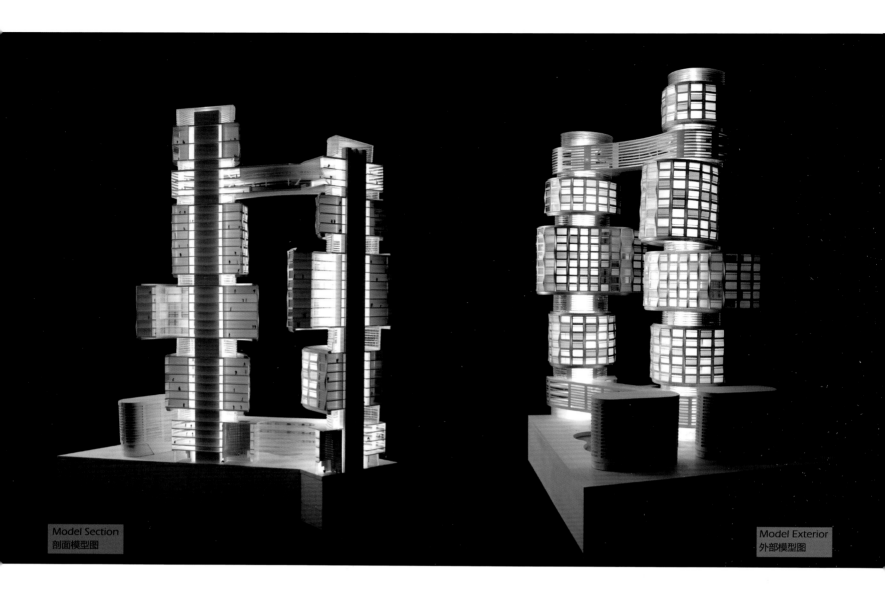

Model Section
剖面模型图

Model Exterior
外部模型图

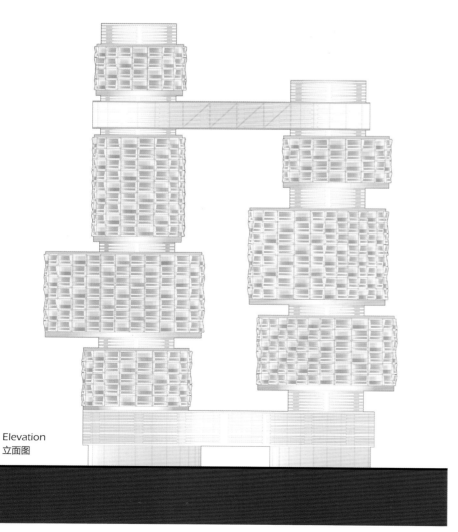

Elevation
立面图

The 500 individual luxury units that vary in size from 45 m² to 82 m² are also designed for compatibility with custom prefabricated plug-in interior components. Asymptote's design of the Velo Towers exploits the latest advances in design, materials and digital fabrication that are now prevalent in present-day automotive, aerospace and marine industries. The merging of these with the latest technological advancements in architecture and the ways in which components can be fabricated and buildings can be assembled, is enabling Asymptote's vision for the Velo Towers to be realized.

500 间个人豪宅单元的面积从 45 平方米到 82 平方米不等，设计也同样考虑到了以后对定制的预制室内部件的兼容性。对于维洛大厦的设计，设计师利用了在设计、材料以及数字制造业中最新的进步手段。这些手段在当今的汽车、航空宇宙和船舶行业中十分盛行。以上元素的融合结合了建筑领域最新的技术手段，以及部件装配和楼体组合的方法，使设计师对维洛大厦的设计愿景得以实现。

FAÇADE

While the overall massing of the Velo Towers is comprised of a dynamic arrangement of rotated and stacked components, the architecture of the towers is further articulated volumetrically and materially at the scale of the façades. The unique faceted façades of the Velo Towers are comprised of large prefabricated components consisting of glass within custom molded composite shells finished in pearlescent automotive paint.

立面

维洛大厦的整体体块是由旋转堆叠的部件以动感的排列形式组成，塔楼建筑在立面比例的处理上是从体积与材质角度进一步铰接的。维洛大厦由各个小平面组成的独特立面皆是以大型预制元组件打造，这些组件是以珍珠色汽车喷涂的定制模型复合外壳的玻璃。

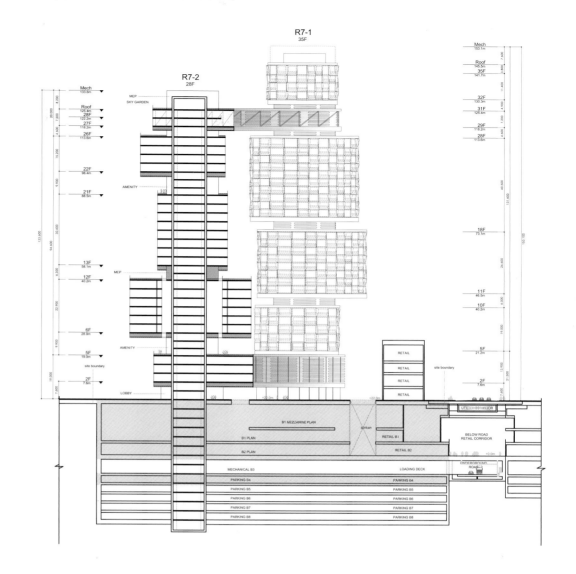

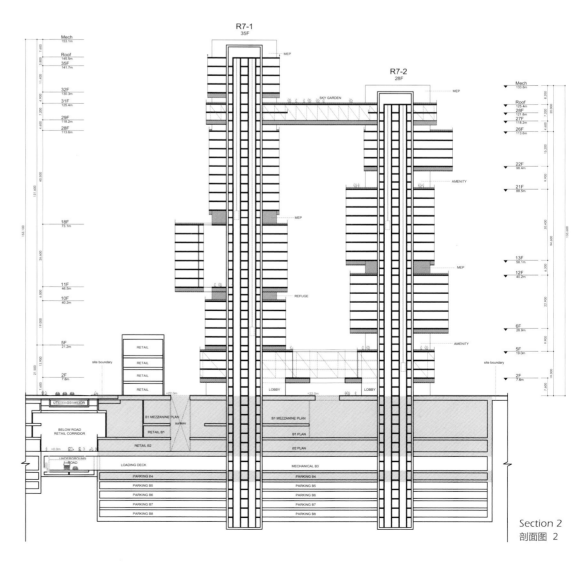

Section 2
剖面图 2

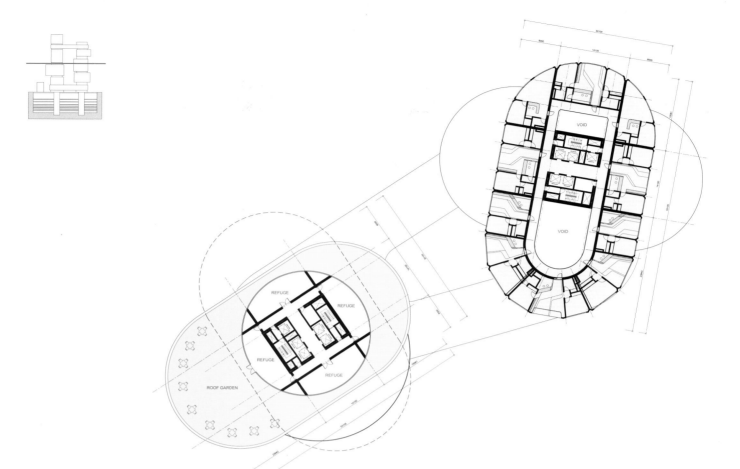

18F Residence Roof Garden Plan
18 层住宅区、屋顶花园平面图

0   5   10   25
SCALE IN METERS

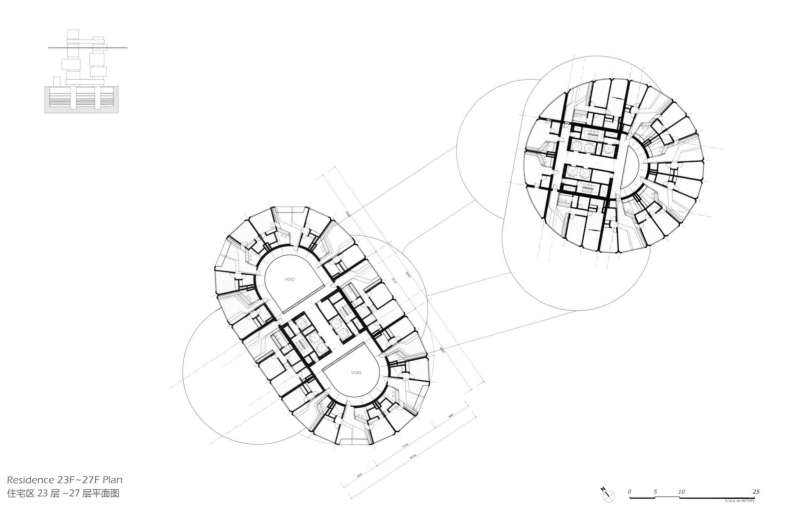

Residence 23F~27F Plan
住宅区 23 层 ~27 层平面图

0   5   10   25
SCALE IN METERS

VIEW 视野

SHAPE 造型

MATERIAL 材料

FAÇADE 立面

Etihad Towers, Abu Dhabi, UAE

阿联酋阿布扎比阿提哈德酒店

Architect: DBI Design
Client: Sheikh Suroor Projects Department (S.S.P.D)
Location: Abu Dhabi, UAE
Gross Floor Area: 500,000 m²
Height: 300 m
Photography: Warren Coyle, Jumeirah, IrfanNaqi, Lasvit

设计公司：DBI Design
客户：Sheikh Suroor 项目署 (S.S.P.D)
地点：阿联酋阿布扎比
建筑面积：500 000 平方米
高度：300 米
摄影：Warren Coyle、Jumeirah、IrfanNaqi、Lasvit

Etihad Towers is a world-leading luxury multi-functional development, consisting of five towers, located on a multi-level podium-base, which is located above four levels of basement car parking. DBI Design orientes the 5 towers, in such a way, as to allow all towers to have unhindered views of the gulf and the surrounding area, including the world's famous Emirates Palace Hotel, and its spacious landscaped grounds. From initial concept design, through design development, and into construction and completion, DBI Design is extremely satisfied with the project's aesthetic and functional outcomes.

阿提哈德酒店是一座世界顶级多功能奢华建筑，整体由 5 座建筑组成，底部是一个多层裙楼区，位于 4 层楼高的地下停车场之上。DBI Design 公司设计 5 座建筑朝向东侧，这样使得建筑可以饱览海湾美景和周围景区，包括世界著名酋长国宫殿酒店及它大面积的优胜景观。从最初的理念设计，到设计开发，再到施工与完工，DBI Design 公司对该项目的美学和功能成果十分满意。

DBI Design employs curves in both the vertical plane and the horizontal plane (in terms of the tower floor-plates) to reduce the towers' visual footprint. (The viewer rarely reads the towers side-on, but always views the towers, as a collection of forms, in perspective.) Etihad Towers, as a series of towers, which spiral above the podium as a consequence of their change of height, creates a complex visual experience of solid and void, light and density, movement and stasis, unparalleled in the world.

DBI 设计在垂直面和水平面中（以建筑楼板为依据）都引入了曲线概念，从视觉上减少其占地面积。参观者很少从一个侧面来解读塔楼，更多的是从远处将其作为一个整体结构来观察。项目作为一个系列型建筑，在高度变化的驱使下以螺旋形状盘旋在裙楼之上，突破低层原有高度，创建一个错综复杂的综合视觉体验，立体又空灵、轻盈又厚重、变化又持久，在世界上无与伦比。

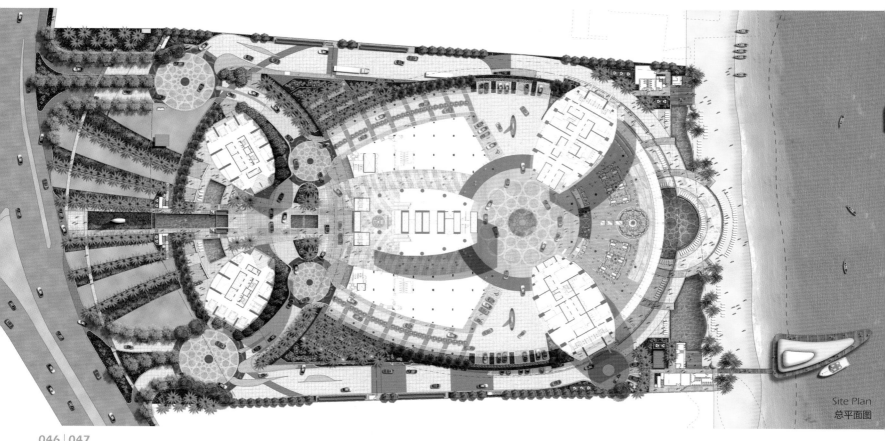

Site Plan
总平面图

FEATURE 特点分析

SHAPE

The appearance of the towers is based upon numerous external visual references, which ranges from the white billowing sails of the traditional Arab dhows, to the curve of the blade of the traditional Arabic sword, to the shape and form of an opening blossom. The tower façade treatment was seen as the need to reflect the elegance of the white sails of the dhow, and the luxury of the lustre found within a pearl shell.

造型

建筑外观的设计引入了诸多外部视觉元素，从传统阿拉伯单桅帆船的白色的波浪型风帆到传统阿拉伯刀剑刀刃般的曲线，再到盛开花朵的形状和构造。建筑立面反映出白帆的优美典雅，以及珍珠贝壳的光泽般的奢华感。

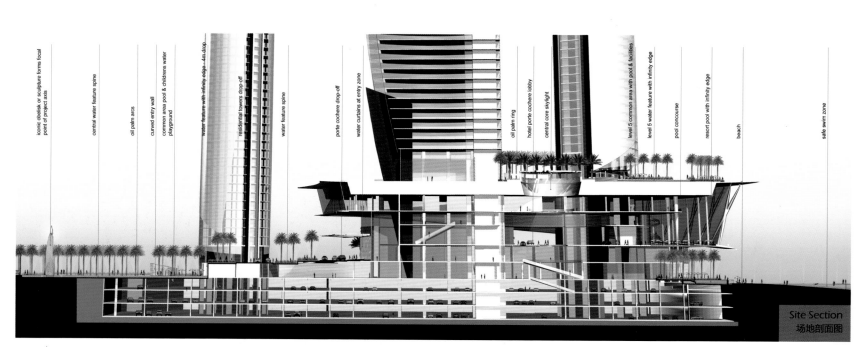

iconic obelisk or sculpture forms focal point of project axis

central water feature spine

oil palm arcs

curved entry wall

common area pool & childrens water playground

water feature with infinity edge - 4m drop

residential towers drop-off

water feature spine

porte cochere drop-off

water curtains at entry zone

oil palm ring

hotel porte cochere lobby

central core skylight

level 5 common area with pool & facilities

level 5 water feature with infinity edge

pool concourse

resort pool with infinity edge

beach

safe swim zone

Site Section
场地剖面图

There is a wide variety of palettes throughout the public spaces to welcome and entice the guests. From airy, light and lofty lobby space, to richly toned pre-function spaces, to the opulent golden ballroom flooded with the lights of bespoke chandeliers, all allow these front of house spaces to undergo dramatic transformations, creating a rich journey as the guests move through the built form.

The rarest of natural stones, finest of European chandeliers and fabrics, hand-tufted rugs all complete the interiors of the public spaces. Each space contains a story of unique material sourced from exotic parts of the world and is hand-selected by the project team. Combined with these are luxurious, contemporary furniture settings from a number of leading Italian manufacturers.

变化的色调贯穿整个公共空间以欢迎和吸引外来游客。从通风明亮高耸的大堂空间到色调丰富的前厅空间，再到漫射着定制吊灯灯光的华丽金色舞厅，房屋空间前部呈现出戏剧性的转变，使客人穿梭于整栋建筑之中时有着丰富的旅行体验。

珍贵的天然石、最高级的欧洲吊灯和纤维织物、工艺地毯填充了整个公共空间的内部。其中，每一个空间设计涵盖了一个源自于异国风情的独有材料的故事，通过设计团队的精心编排而最终形成。结合这些元素，设计师还选用了意大利顶级制造商生产的奢华、当代的家居布置。

Structural Analysis
结构分析图

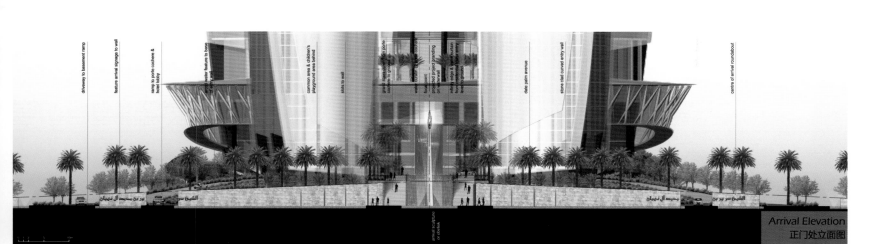

Arrival Elevation
正门处立面图

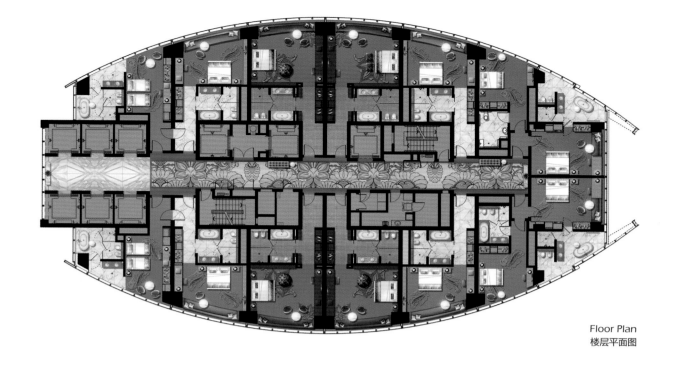

Floor Plan
楼层平面图

The interior design for accommodation in Etihad Towers complements the sleek, iconic form of the architecture.

The inspiration is drawn from the fusion of Middle Eastern motifs and contemporary lifestyle and aspirations for the future. The accommodation in the hotel tower celebrates the modernity of Etihad Towers and wraps the guest in the exotic opulence of the Middle East. The rooms have been designed to provide a range of guest experiences: light colour and neutral palette of colours are contrasted with rich decorative elements. Colour, pattern and material choice have been extremely critical points in creating a unique design vision for this project. The stone choice is very unique to the project, and adds another level of sophistication and luxuriousness to the design.

Sumptuous carpet patterns have all been custom designed by DBI Design for this unique hotel, with abstract design patterning combined with Middle Eastern traditional decorative motifs. In addition, all rooms have been designed to use smart integrated technology. Taps and shower heads comply with stringent water use efficiency.

阿提哈德酒店的客房内部设计犹如建筑圆滑形象造型的补充。

设计灵感源自于中东的装饰图案与现代生活格调以及对未来生活的期望的融合。酒店内住宅设施反映了该建筑的现代时尚感,将客人包围在丰富的中东异国情调中。房间设计为客人提供一系列体验:浅色和中性色调与丰富的装饰性元素形成了鲜明的对比。色彩、图案和材质的选择成为这一项目实现的关键性因素。石材的选择对于项目来说也是非常独特的,并且为设计增添了更多的完善和奢华感。

华丽的地毯图案是 DBI Design 公司为打造独一无二的酒店而定制的,以抽象的模式结合中东传统装饰图案。另外,所有的房间设计都应用智能集成技术。水龙头、淋浴喷头都是严格遵守水分利用效率来设计的。

SUSTAINABILITY 可持续性

SHAPE 造型

STRUCTURE 结构

FAÇADE 立面

Bratislava Culenova New City Centre

斯洛伐克布拉迪斯拉发库雷诺瓦新城中心

Architect: Zaha Hadid Archtiects
Client: Penta Investments Limited o.z.
Location: Bratislava, Slovakia
Gross Floor Area: 150,000 m²
Height: 115 m

设计公司：扎哈·哈迪德建筑师事务所
客户：Penta 投资公司
地点：斯洛伐克布拉迪斯拉发市
建筑面积：150 000 平方米
高度：115 米

The scheme is based on a dynamic field strategy which organizes the new city centre's program along a gradient of circular and elliptical patterns.

项目规划基于动态场地规划策略，通过圆形和椭圆形的渐变模式来对新城市中心进行组织规划。

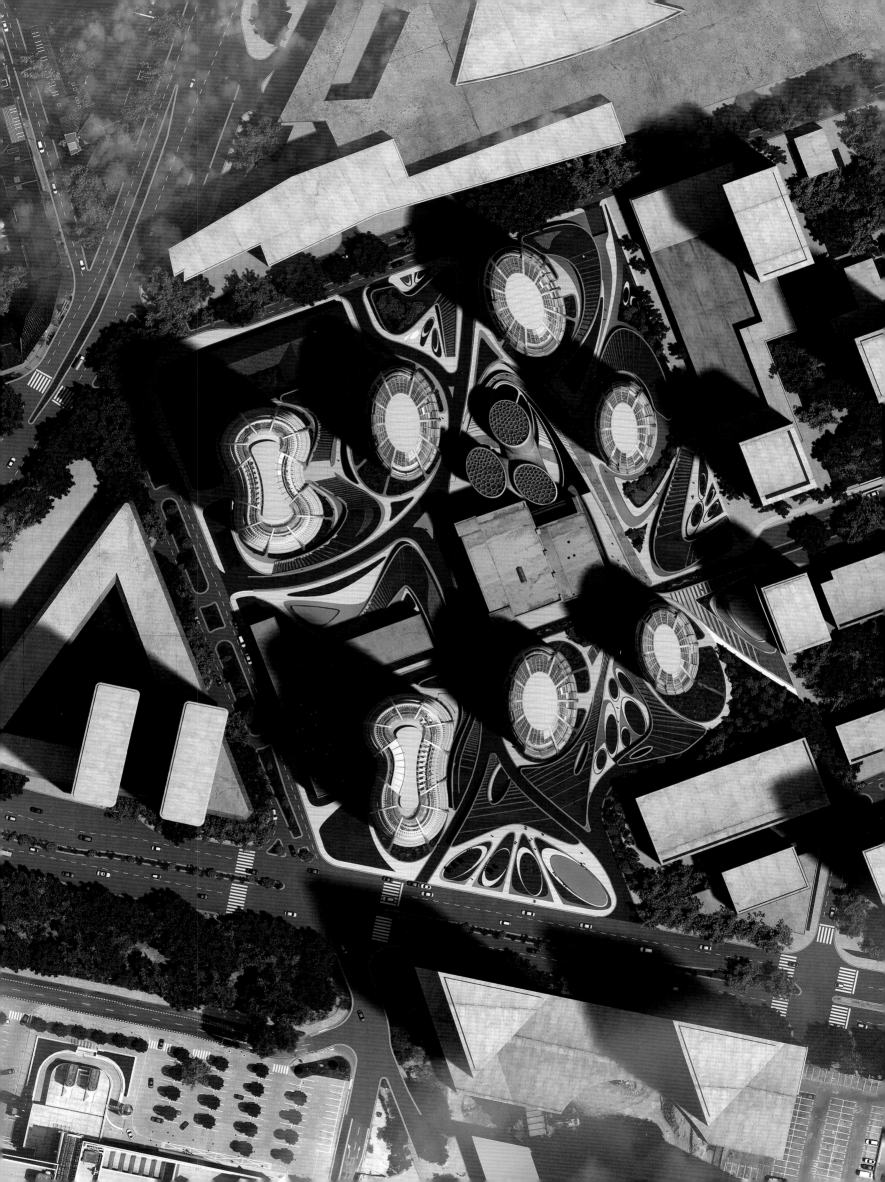

A fluid field emerges from the underlying matrix in a series of larger tower extrusions towards the site's perimeter and intermediate scale pavilion-like structures surrounding the cultural plaza next to the existing power plant.

In order to activate the ground throughout the whole site and provide high-quality public space, the double-storey underground car park is covered by a one-storey high modulated slab, which is perforated at strategic points for day-lit patio spaces in order to accommodate retail and public landscape with various points of interest such as the cultural centre with its museum shop and further public attractors (e. g. conference space and event halls).

Towards the site's perimeter, the slab is slightly lifted up at specific points in order to define the site's edge and accommodate programmatic points of interest, access points to the parking level and access to office and housing related program. While at other strategically chosen areas, it merges with the surrounding city level to create a highly blurred site's boundary condition that links the new urban parks and plazas with the surrounding city fabric.

The scheme creates high density via efficient high-rise structures, providing a generous and highly activated ground level with public spaces that are gradually differentiated within a three-dimensional field condition.

一个流动的场地建于底层的矩阵中,周边分布有朝向场地外围的多座高塔与围绕着紧邻现有发电厂的中等尺度的亭状结构。

为了充分利用场地的地面空间,并打造出高品质的公共空间,两层高的地下停车场的顶部覆盖了一层楼高的调制板,在调制板的战略点上进行穿孔,是为了保证中庭空间可受到日光照射,以适应不同的特色的零售区与公共风景区,如包括博物馆商店和稍远的公共空间(如会议室和活动大厅)在内的文化中心。

场地的边缘在特殊点略微抬高,以划定场地边界并容纳人文景观、停车场出入口及办公和住宅相关区域的规划。在另外一些特定区域,它和周边环境融合在一起,模糊了彼此之间的界限,将新城和广场与周边城市脉络相连。

该方案通过高效的高层结构创建了一种高密度形式,同时形成了宽敞又灵活的地面公共空间,这些空间随着立体地面条件变换呈现不同的效果。

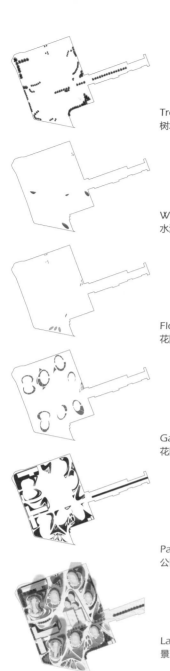

Trees
树木分布图

Water
水源分布图

FlowerBed
花圃分布图

Garden
花园分布图

Parks/Lawn
公园 / 草坪分布图

Landscape Planning
景观规划图

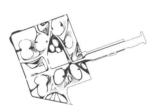

Secondary Paths
副通道图

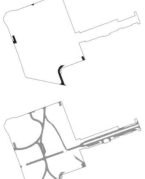

Car park Access
停车场入口图

Main Paths
主通道图

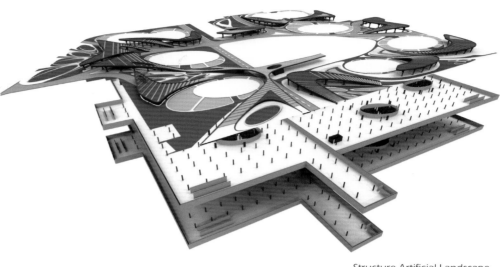

Structure Artificial Landscape
人造景观结构图

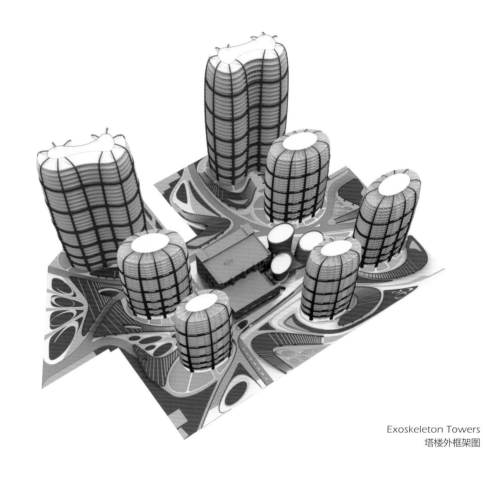

Exoskeleton Towers
塔楼外框架图

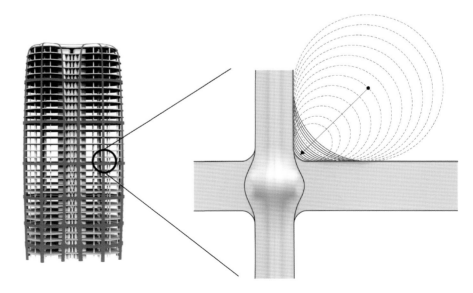

Structural Detail Exoskeleton
外框架结构节点图

FEATURE 特点分析

SHAPE

Cylindrical shape of the single architecture fades down the edge of building volumes, broadening the views between the buildings and thereby lowing the high density the buildings were in.

造型

单体建筑的圆筒型造型淡化了建筑体量的边缘，开阔了楼与楼之间的视野，从而降低了楼群间的高密度。

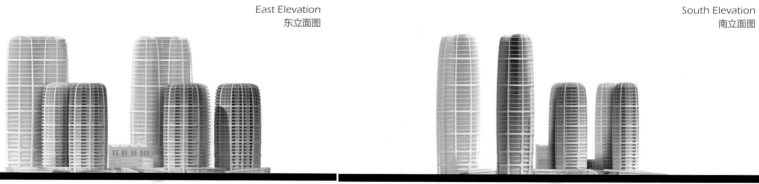

East Elevation
东立面图

South Elevation
南立面图

West Elevation
西立面图

North Elevation
北立面图

Section AA
剖面图 AA

Section BB
剖面图 BB

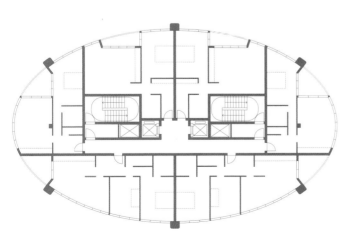

Floor Plan 1
楼层平面图 1

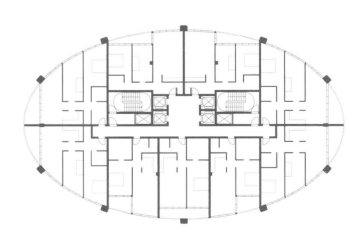

Floor Plan 2
楼层平面图 2

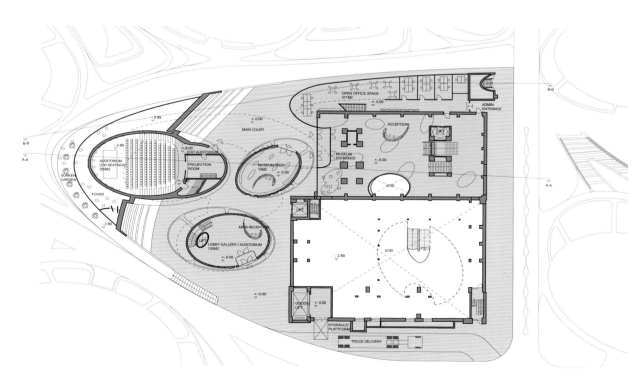

Museum Plan
展览馆平面图

SHAPE 造型

ECOLOGY 生态

MATERIAL 材料

ENERGY SAVING 节能

Sanya Phoenix Island, Sanya, China

中国三亚凤凰岛

Architect: Jiang Architects & Engineers, MAD
Location: Sanya, China
Gross Floor Area: 460,000 m²

设计公司：上海江欢成建筑设计有限公司、MAD 建筑事务所
地点：中国三亚市
建筑面积：460 000 平方米

Phoenix Island, located in Sanya Bay of Sanya, is an artificial island with the total land use area of 3,650,000 m², and is the Sanya international mail steamer harbor. The total building floor area of the island is about 500,000 m², including international conference centre, international health preserving and vacation centre, business club, international cruiser club, Olympic square, commercial streets and other projects.

凤凰岛位于三亚市三亚湾，为总土地使用面积 365 万平方米的人工岛，是三亚国际邮轮港所在地。全岛总建筑面积约 50 万平方米，由国际会议中心、国际养生度假中心、商务会所、国际游艇会所、奥运广场、商业街等多个项目组成。

Relying on Sanya Bay, taking full advantage of the seashore, the Lu Huitou scenery spot and other fine natural landscapes, following the principle of sustainable development of society, economy, culture, environment and technology, and laying stress on the principle of people-orientation, this project seeks to form an international seashore vacation urban landscape that is closely connected to its surroundings. The three-dimensional annular 200 m-high international conference centre and the 5 hotels over 100 m-high international health preserving and vacation centre, and the dual-curved servicing apartments have become new landmarks in Sanya.

设计依托三亚湾，充分利用海岸、鹿回头景区及其他良好的自然人文景观的优势，遵循社会、经济、文化、环境和技术的可持续发展原则，强调以人为本，力求形成与周边环境结合紧密的国际海滨休闲度假城市景观。高达 200 米的立体环状国际会议中心及酒店、5 栋 100 多米高的国际养生度假中心以及双曲面形态的酒店式公寓，成为三亚新的地标。

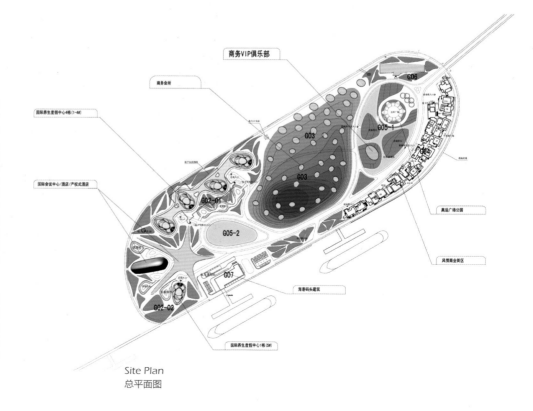

Site Plan
总平面图

Ground Floor Plan
首层平面图

FEATURE 特点分析

SHAPE

The shape is fashionable and dynamic, which has a full consideration on the harmonious relationship between the environment, architecture, space, time and human. Thus, this building space becomes the new landmark in the region through creating a modern atmosphere and blending with the environment.

造型

时尚、灵动的非线型造型充分考虑了环考虑了环境、建筑、空间、时间以及人的和谐关系。营造成既前卫又能溶入环境的建筑空间，成为地区的新地标。

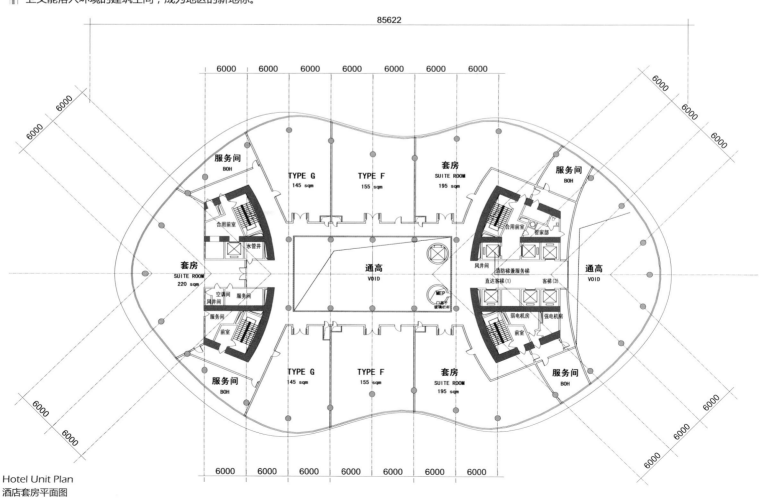

Hotel Unit Plan
酒店套房平面图

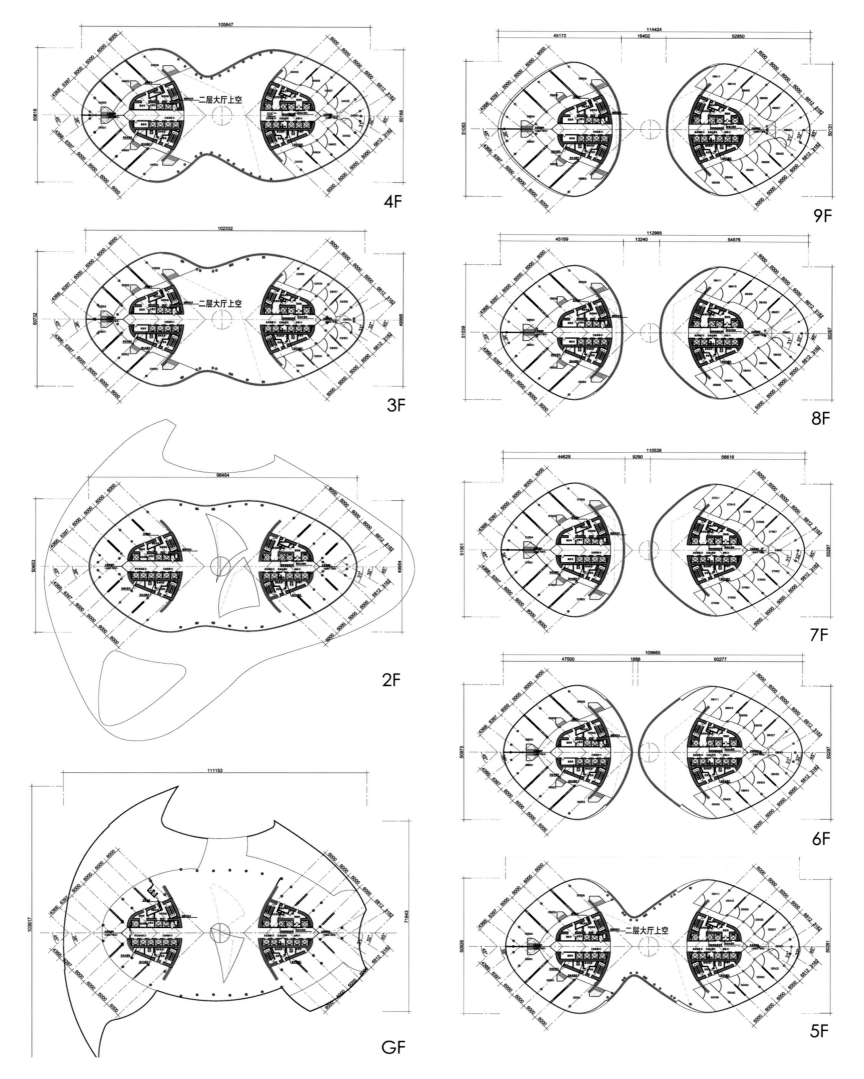

4F

9F

3F

8F

2F

7F

6F

GF

5F

Floor Plan
楼层图

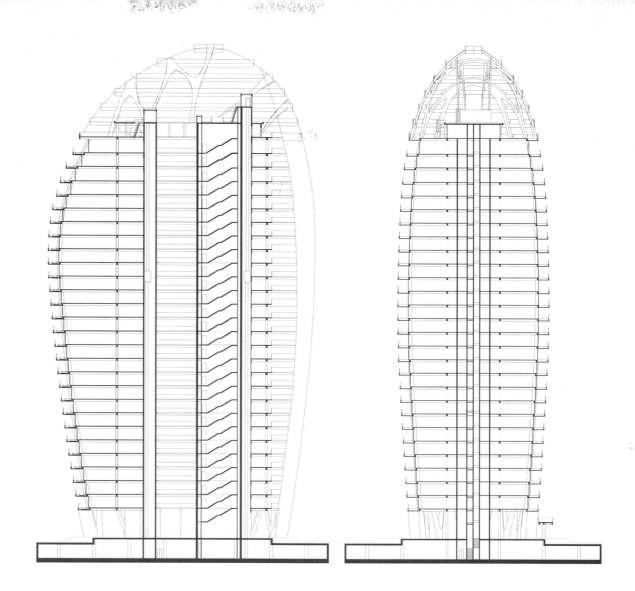

Section
剖面图

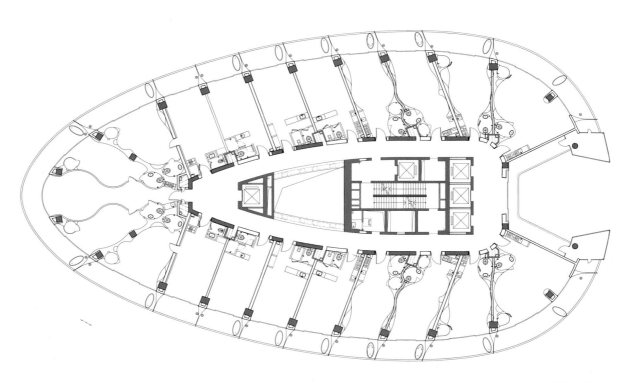

Standard Floor Plan
标准层平面图

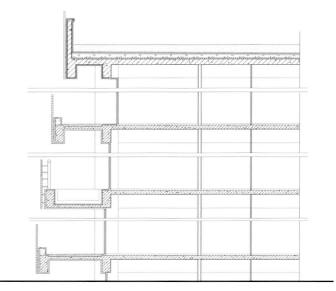

Exterior Wall Detail
外墙局部节点图

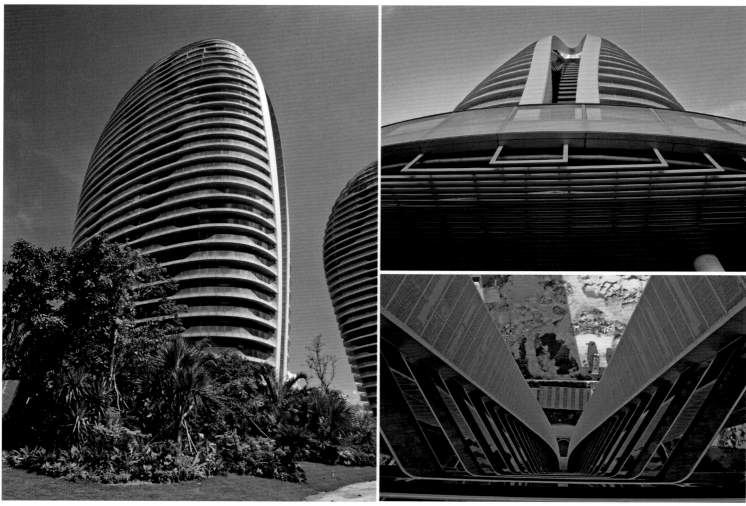

FAÇADE 立面

SHAPE 造型

STRUCTURE 结构

VIEW 视野

London Bridge Tower (The Shard), London, UK

英国伦敦桥大厦（夏德大厦）

Architect: Renzo Piano Building Workshop, Adamson Associates
Client: Sellar Property Group
Location: Southwark, London, UK
Site Area: 110,000 m²
Gross Floor Area: 126,712 m²
Floors: 87
Height: 310 m
Photography: Rob Telford, Potto-Skyscraper, Michel Denancé, Renzo Piano Building Workshop, Paul Raftery
Realization: 2012

设计公司：伦佐·皮亚诺建筑工作室、亚当森联合公司
客户：Sellar 物业集团
地点：英国伦敦市萨瑟克区
占地面积：110 000 平方米
建筑面积：126 712 平方米
层数：87 层
高度：310 米
摄影：Rob Telford、Potto-Skyscraper、Michel Denancé、伦佐·皮亚诺建筑工作室、Paul Raftery
完成时间：2012 年

London Bridge Tower, which is also known as the Shard, is a 72-storey mixed-use tower located besides London Bridge Station on the south bank of the river Thames. The station, which combines train, bus and underground lines, is one of the busiest stations in London with 200,000 users per day. The project is a response to the Mayor's policy of promoting high density development at key transport nodes.

伦敦桥大厦，也称"夏德大厦"，是一栋 72 层高的综合型大楼，位于泰晤士河南岸的伦敦桥站附近。伦敦桥站是伦敦最繁忙的车站之一，集火车、公交和地铁服务于一体，每天客流量高达 20 万人。该项目是为了推动关键运输点高密度发展的市政策而建立的。

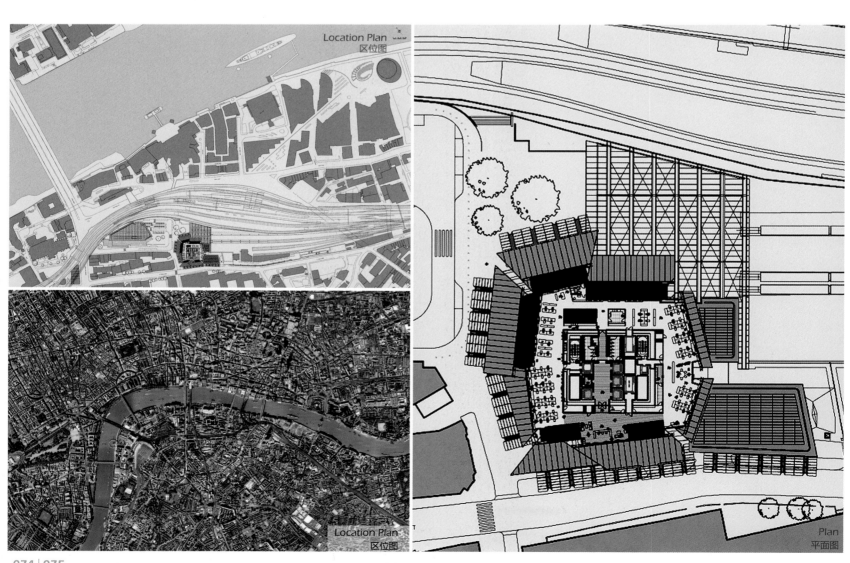

Location Plan
区位图

Location Plan
区位图

Plan
平面图

The 310 m-high London Bridge Tower is one of the tallest buildings in Europe, also a landmark in London. Slender spire model, advanced materials, rectangular exterior panes, all are designed to reflect the change of weather and seasons as the outlook of building will also be changed. The form of the tower is determined by its prominence on the London skyline. Unlike other cities such as New York or Hong Kong, the London Bridge Tower is not part of an existing cluster of high-rise buildings. References include the masts of ships docked in the nearby Pool of London and Monet's paintings of The Houses of Parliament.

The project also includes the redevelopment of the train station concourse and bus station. The existing roof is to be removed and replaced with a glazed canopy, and retail units are relocated to open up visual connections between the train station, bus station and taxi ranks. Two new 30 m x 30 m public squares will form the centre of the scheme. Such improvements to the public realm are vital to the regeneration of this congested and neglected part of the city and will hopefully provide the catalyst to further redevelopment in the area.

高达 310 米的伦敦桥大厦是欧洲最高的建筑物之一，也是伦敦市的新地标。细长的尖顶造型、先进材料的采用、直角外墙玻璃窗格，建筑物的外观随着天气和季节的变换也将改变。建筑的造型设计取决于它在伦敦天际线上的突出地位，不同于纽约或香港之类的城市，它并非现有高层建筑群的一部分。项目造型设计参考了基地附近伦敦港船舶桅杆以及莫奈的画作《议会大厦》。

该项目还包括了对火车站广场和汽车站的重建。现有的顶部将被拆除，代之以玻璃天棚，商业设施也将搬离，以打开火车站、汽车站和出租车停靠站之间的视野。而项目中心将由两个新的长宽各 30 米的广场组成。这种对于公共空间的改善将会成为这个拥挤又被忽视的城市空间里的新生契机，也有望成为这个区域长远发展的催化剂。

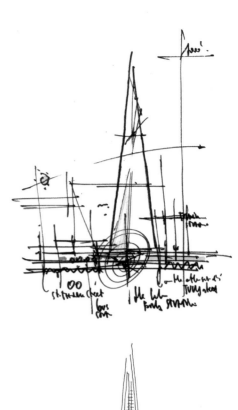

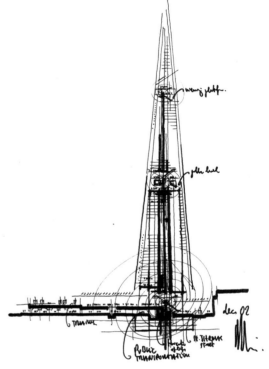

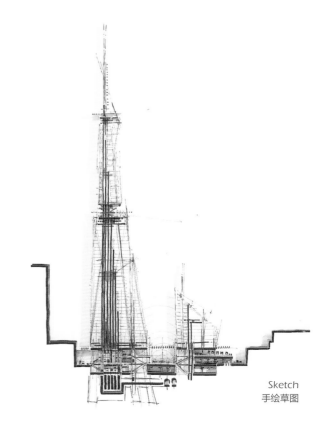

Sketch
手绘草图

Eight glass shards define the shape and visual quality of the tower. The passive double façade uses low-iron glass throughout, with a mechanised roller blind in the cavity providing solar shading. In the "fractures" between the shards, opening vents provide natural ventilation to winter gardens. These can be used in meeting rooms or break-out spaces in the offices and winter gardens on the residential floors. They provide a vital link with the external environment often denied in hermetically sealed buildings.

The main structural element is the slip-formed concrete core in the centre of the building. It houses the main service risers, lifts and escape stairs. A total of 44 single and double-deck lifts link the key functions with the various entrances at street and station concourse level.

8 片玻璃幕墙决定了建筑外形和大厦的视觉质量。双层被动式幕墙全部采用了低铁玻璃，幕墙内的凹槽安装了机械滚动百叶，能起到遮阳作用。通风口为冬季花园提供了自然通风，这种通风口被用在办公层的会议室或休息空间以及住宅层的冬季花园中，打破了原有的封闭式建筑结构，让建筑与外部形成了良好的互动。

建筑的主要结构是位于其中心的滑模混凝土芯，承载了主要的管道、升降梯和消防通道。单层和双层电梯总计 44 台，将各功能区同位于首层的各出入口和车站大厅联系起来。

FEATURE 特点分析

SHAPE

The architectural shape refers to the masts on the London Port, which appeared in a conical form enclosed by 8 glass curtain walls. It blurs the skyline as disappearing in the sky gradually to give a better integration with the surroundings.

造型

建筑的造型参考了伦敦港的桅杆，它是 8 块玻璃幕墙围合成的锥形造型，由下而上消失在天空中，使城市天际线不那么突兀，让建筑更好地融入周围的建筑环境。

General Section
剖面图

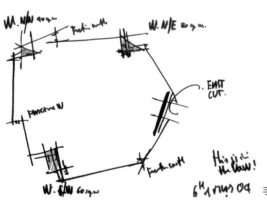

Sketch
手绘草图

L39 Plan
39 层平面

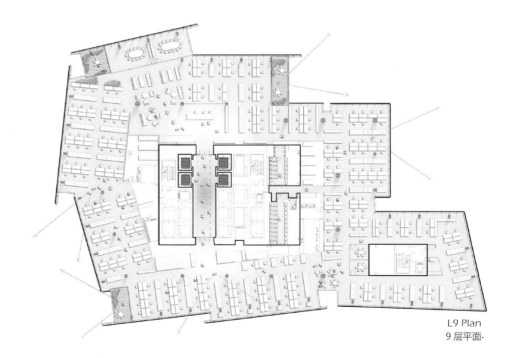

L9 Plan
9 层平面·

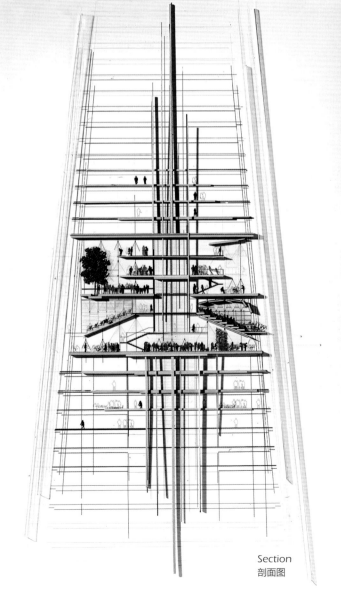

Section
剖面图

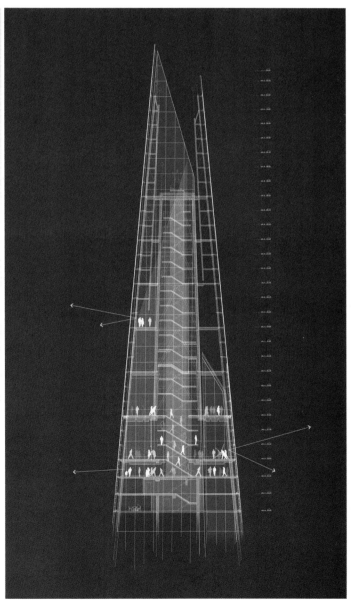

MATERIAL 材料
ECOLOGY 生态
SHAPE 造型
FAÇADE 立面

South Beach, Singapore
新加坡南海滩综合体

Architect: Foster + Partners, Aedas
Client: South Beach Consortium Pte Ltd
Location: Singapore
Gross Floor Area: 150,000 m²
Realization: 2014
Images: South Beach Consortium Pte Ltd, Arup

设计公司：Foster + Partners, Aedas
客户：South Beach Consortium Pte Ltd
地点：新加坡
建筑面积：150 000 平方米
完成时间：2014 年
图片：South Beach Consortium Pte Ltd、奥雅纳

Occupying an entire city block between the Marina Centre and the Civic District, the scheme will create a 150,000 m² eco-quarter in downtown of Singapore that continues the Singaporean ideal of the "city in a garden" with its lush planting and sky gardens.

The scheme incorporates commercial, residential, retail and two high-end hotels, as well as a direct "green" link to an MRT station. Offering a light and comfortable environment, Beach Road will provide an exemplar sustainable quarter for Singapore.

项目位于滨海中心与市政区中间的一整片城市街区，在新加坡市中心区形成了一处 150 000 平方米的生态区，它继承了新加坡意图打造伴有繁茂苍翠与空中花园的"花园中的城市"的理念。

项目包含商业、住宅、零售和两个高端酒店等各种用途区，并设"绿色"通道直通附近的地铁站。它将成为本地可持续区规划的一个范例，给人们提供轻松、舒适的环境。

FEATURE 特点分析

ECOLOGY

Central feature is a ribbon-like and lightweight canopy that is inspired by the natural form and function of a tree. The 280 m-long canopy floats across the entire development to filter sunlight, solar glare and channel wind breezes into the public spaces. The result is a comfortable and cool environment all year round. At the same time, lower points of the canopy create valleys where rainwater can be collected and recycled for irrigating

The generous canopy protects the public realm at ground level, buffering the spaces beneath from the extremes of the tropical climate. Above the canopy rises a vertical city of clustered towers. Green features include extensive sky gardens which like the canopy, act as "lungs" to create greater natural air movement. These innovative features will contribute significantly to minimise the development's overall energy consumption. In fact, it is estimated that South Beach will save close to 2,000 MWh of electricity and 174,000 m³ of water annually.

FAÇADE 立面

STRUCTURE 结构

SHAPE 造型

QUAKE-PROOF 减震

Dorobanti Tower, Bucharest, Romania

罗马尼亚布加勒斯特多罗班蒂大厦

Architect: Zaha Hadid Architects
Client: Smartown Investments
Location: Bucharest, Romania
Site Area: 10,000 m²
Gross Floor Area: 100,000 m²
Height: 200 m

设计公司：扎哈·哈迪德建筑师事务所
客户：Smartown Investments
地点：罗马尼亚布加勒斯特市
占地面积：10 000 平方米
建筑面积：100 000 平方米
高度：200 米

The brief calls for a 100,000 m² mixed-use, high-rise development in the centre of Bucharest. The project comprises a five-star hotel, luxury apartments, retail areas, a business centre, a casino and underground parking.

The Dorobanti Tower is designed to establish an iconic presence in the heart of Bucharest. The new tower is a unique mix of a distinctive form, ingenious structure, and spatial qualities of sky-high living. The purity of its form — a chamfered diamond-like structure — will be a new landmark in the centre of Bucharest. Zaha Hadid Architect's design concept is a synthesis of architecture and engineering, which integrates a distinct meandering structural lattice.

在布加勒斯特市中心开发的这个项目要求建造一座占地 10 万平方米混合功能的超高层建筑，它包括一家五星级酒店、超豪华公寓、商业零售区、商务中心、俱乐部和地下停车场。

多罗班蒂大厦被设计成布加勒斯特市中心地区的标志。新塔楼是一座集独特形式、独创结构和奢华生活的空间品质于一身的独特建筑。它特别纯净的造型——倒棱的钻石状结构，将成为布加勒斯特市中心的新地标。扎哈·哈迪德建筑师事务所的理念是：塔楼应该是一个建筑风格和工程学的综合体，是一个有着独特曲线的结构晶体。

Bucharest is within a vulnerable seismic zone and therefore the structural concept was crucial to us from the beginning of the design process. Placing the primary structure at the exterior not only maximises the structural footprint but also allows for column-free interior spaces.

Concrete-filled stainless steel profiles follow in sinus waves from the ground level to the top of the tower, creating a distinctive identity, complementing the tower design. The concrete's filling will give additional strength to the structure and it will provide fire protection to the steel profiles. The façade structure expresses the various programmes and it adjusts from bottom to top according to the changing structural forces. The secondary structure, which is integrated between the main steel tubes, gives additional strength to the primary steel mesh and it also works as a damper in case of earthquakes.

The public realm adjacent to the tower will be unlike anything else in Bucharest, representing a major attraction within the dense urban character of the city, offering an important new meeting space and urban plaza. Metaphorically speaking, the landscaping can be understood as a warped concrete "carpet" with one continuous surface connecting the three surrounding streets. The landscape is sculpted to create seating areas, water basins, fountains, green areas including trees and a raised terrace.

Urban parameters, site constraints and the programme generate the building's elegant tapering profile. The new tower establishes a distinctive identity while avoiding sterile repetition through its dynamically changing profile. In order to maximise lighting and views for the neighbourhoods, the elongated curvilinear shape reduces its perimeter towards the top, while the offset to the inside at ground floor level shall create a generous public realm and an appropriate entrance plaza in front of the tower.

布加勒斯特处在易受损的地震区，因此结构的观念从设计一开始就显得非常重要。塔楼表面按一级结构布置，不但将结构封装最大化，还形成了塔楼无柱的内部宽敞空间。

填充混凝土的不锈钢直径随着塔楼从底层到顶部的起伏变化而变化，生成了独有的特性，使设计更加完美。混凝土的浇入加强了结构强度，也对钢筋提供了防火保护。立面的构造表达了多种方案，它根据结构应力的变化，从底部到顶部进行调整。二级结构布置是钢管之间的整合，它增强了通用钢网的强度，发生地震时能起到减振作用。

毗邻塔楼的公共区域将一点也不同于布加勒斯特的其他地方。在密集的城市特征的市区里，它标志着一个巨大的吸引力，提供了一处重要的新集会空间和城市广场。环境美化工程，可看作是一条弯曲的混凝土"地毯"，连续的表面连接着周围的三条街道。景观带被切分出座位场地、水池、喷泉、绿化区域，这些绿化区域种植了很多树木并有一座架高露台。

城市参数、场地制约和规划促成了建筑物优雅的锥状外形。新塔楼通过它动态变化的外形避免了乏味的重复，建立了有独特风格的特征。为了使邻域得到最大程度的采光和视野，它细长的曲线形状缩小了朝向顶部方向的周长，同时为了补偿内部空间，在底层将创建一个大的公共区域，在塔楼的正面创造一个合适的入口广场。

200.00

184.00 Penthouse

105.80 Residential
102.40 Bar
99.00 Spa
95.60 Pool
92.20 Technical

28.00 Hotel

20.00 Conference Centre

12.00 Restaurant

00.00 Lobbies/ Retail

-17.20 Parking/ Plantroom

Section
剖面图

FEATURE 特点分析

STRUCTURE

At the bottom, the façade grid has denser amplitudes according to the structural requirements for a tower of this height, providing the required load-bearing capacity and stiffness to the structure. At the technical and recreation levels, the structure condenses, creating almost solid knots. On the top floors, the primary structure has been reduced to the minimum according to the structural requirements in order to maximise the views and lighting for the luxury apartments.

结构

在塔楼的底部，根据塔楼高度所需的结构要求，表面网格密集，以便为建筑物提供足够的负荷能力和抗挠性能。在设备、娱乐层，表面结构几乎连成一个整体，而在顶层，在满足结构要求条件下，一级结构减至最低值，豪华公寓的观光和照明得以最大化满足。

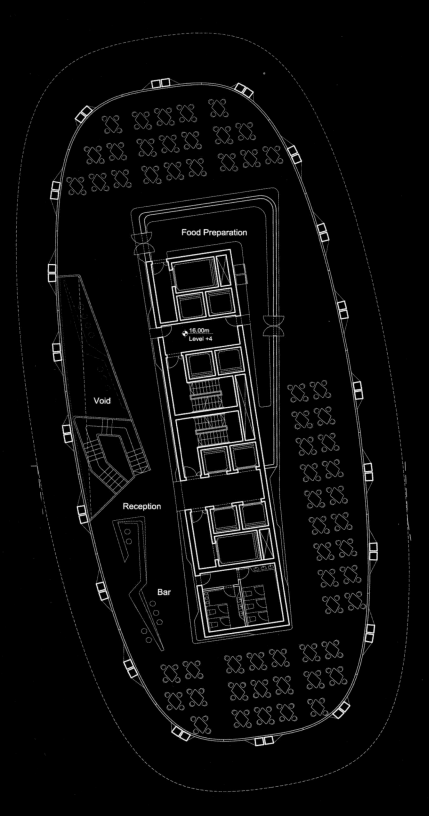

Food Preparation

16.00m
Level +4

Void

Reception

Bar

4th Floor Restaurant Plan
4 层餐厅平面图

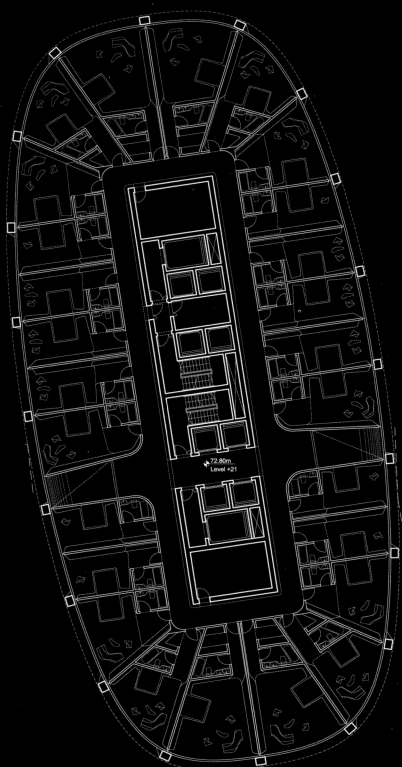

72.80m
Level +21

Typical Hotel Floor Plan
酒店层典型平面图

088 | 089

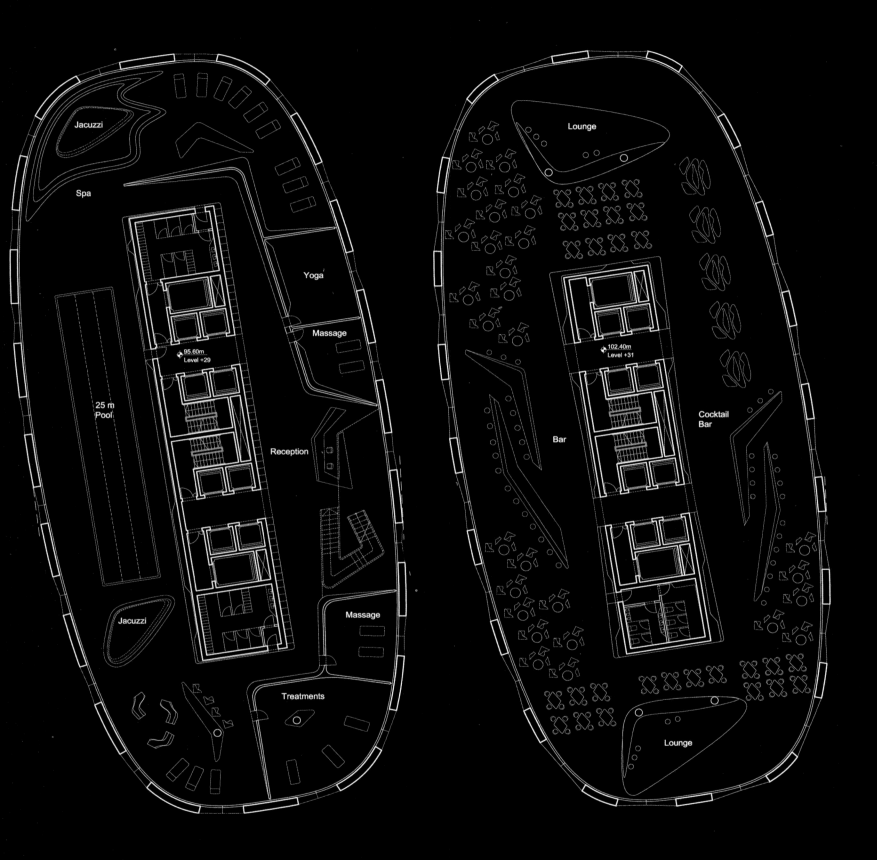

Jacuzzi

Spa

Yoga

Massage

+ 95.60m
Level +29

**25 m
Pool**

Reception

Jacuzzi

Massage

Treatments

Lounge

+ 102.40m
Level +31

Bar

**Cocktail
Bar**

Lounge

MATERIAL 材料

SHAPE 造型

FAÇADE 立面

ENERGY SAVING 节能

Wu Zi Apartment, Shenzhen, China

中国深圳伍兹公寓

Architect: Steffian Bradley Architects,
The Architecture Design & Research Institute of Guangdong
Province Branch of Shenzhen
Client: Shenzhen Merchants Property Development Co., Ltd.
Location: Shenzhen, China
Site Area: 11,319.49 m²
Gross Floor Area: 47,211.4 m ²
Floors: 28
Photography: Shenzhen Pioneer Space Media Co., Ltd.

设计公司：美国 SBA 建筑设计有限公司、
广东省建筑设计研究院深圳分院
客户：深圳招商房地产有限公司
地点：中国深圳市
占地面积：11 319.49 平方米
建筑面积：47 211.4 平方米
层数：28
摄影：深圳先锋空间文化传媒有限公司

Wu Zi Apartment is located at the Sea World, Nanshan District, Shenzhen City, which has a built area of 100,000 m² and will become an international coastal recreation town including restaurants, entertainment facilities, shopping, hotels, office buildings, art galleries, leisure places and residential buildings. Wu Zi Apartment will be the first new community, located in the heart of Sea World, enjoying an open seaside landscape. LOW-E hollow glass curtain wall, metal wall, clay plates as main material will create a landmark mansion.

伍兹公寓坐落于深圳市南山区海上世界。海上世界建筑
面积达 10 万平方米，将成为一个集餐饮、娱乐、购物、
酒店、写字楼、艺术馆、休闲场所及住宅于一体的国际
滨海休闲城区。伍兹公寓将是海上世界首个新楼盘，位
于海上世界中心地带，尽揽美景，并运用低辐射中空玻
璃幕墙、金属幕墙、陶土板等主要材料，打造地标式豪宅。

FEATURE 特点分析

SHAPE

The surface texture on the plinth echoes well with upper parts. The solid plinth adopts deeper recess while balconies are overhang on one side of the building. Seen from the sea, three buildings flash at night like three lighthouses standing on the harbor.

造型

建筑基础的表面纹理和上层部分相互衬托。基座坚实浑厚，采用较深的凹进，塔楼主体一侧营造水平外伸的阳台。从海上看，三栋建筑在晚上会发出闪烁的光芒，犹如三座海港上发光的灯塔。

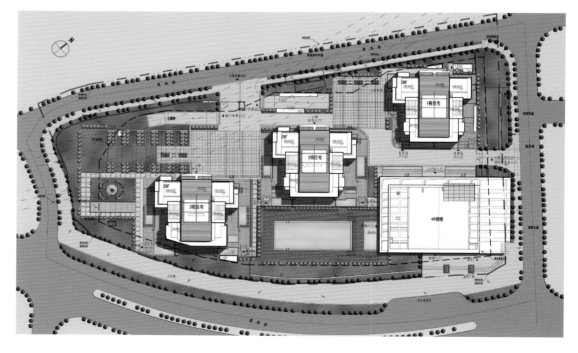

Site Plan
总平面图

The project mainly uses clay plates which have the feature of energy saving and environmental protection, while enjoying an appearance like traditional materials for its exquisite workmanship.

The volume of each tower is divided into four parts around the core. The main façades of each part are covered by a translucent white glass frame with different heights.

The façade design gets inspiration from the traditional Chinese Curio Display Grid. Balconies in different settings and masses, screens and bay windows for privacy protection together form an interesting visual effect.

The commercial volume in the building is equipped with French windows, which are covered by vertical white glass louvres on the outside layout for sun-shading while echoing with the residential volume façade.

The roof and roof slabs on bay windows use 40 thick extruded polystyrene as an insulation layer. Exterior wall uses 200 thick aerated concrete bricks, reinforced concrete exterior walls from east to west use 20 thick thermal mortar and external windows use LOW–E hollow glass. The perimeter structure of the project is perfectly completed that truly reaches green energy-saving standards.

本工程主要采用陶土板，该材料节能环保、工艺精湛，并具备传统土制材料的外观。

每栋塔楼的体量被分为四个围绕核心筒的体块。每个体块的主立面被包围在半透明白色玻璃框架内，具有不同的高度。

立面设计从中国传统的多宝格中获取灵感。通过采用不同模数和体量的阳台，为保护私密性而设置的网屏以及凸窗等元素，形成了趣味性视觉效果的图案。

商业采用大面积落地玻璃，玻璃外侧采用竖向白色玻璃百叶，和住宅立面协调统一，同时达到遮阳的效果。

项目屋面和凸窗顶板均采用 40 厚挤压聚苯板做保温层，外墙采用 200 厚加气混凝土砌块，东西向钢筋混凝土外墙采用 20 厚保温砂浆，外窗采用铝合金低辐射中空玻璃窗。该项目的外围结构很完善，真正达到了绿色节能的标准。

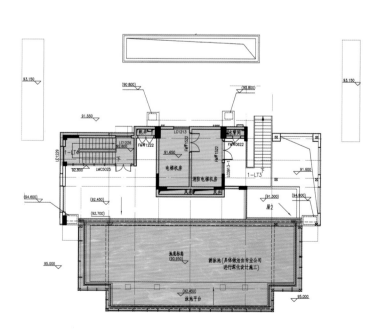

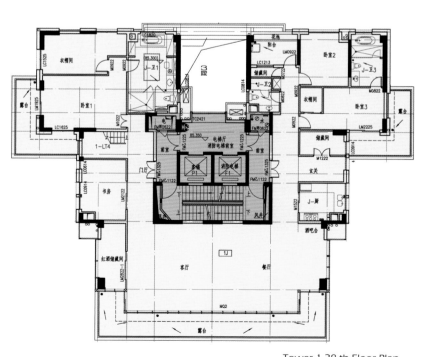

Tower 1 Floor Plan
一栋楼层平面图

Tower 1 28 th Floor Plan
一栋 28 层楼层平面图

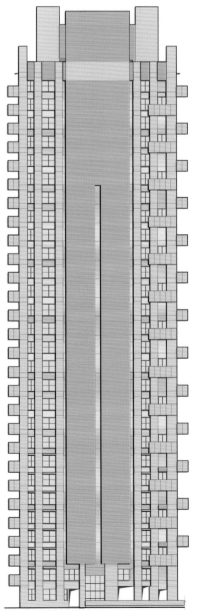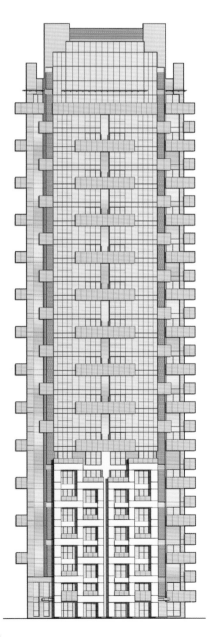

The unite layout is designed into a T-shape, equipping at least one living room or bedroom in each household with an open seaside view on the east. Positioned as high service apartment, the design is to guarantee indoor space open and elegant by a slab structure with fewer pillars and no beam so as to increase the flexibility of space interval.

住宅平面布置成 T 形，使得每户视野开阔，且至少有一个客厅或是卧室能获得东向海景。本项目定位于高级服务性公寓，因此在户型设计上均保证了室内空间开敞和大气，并采用了大板结构，使得室内空间无梁少柱，增加了空间间隔的灵活性。

Elevation
立面图

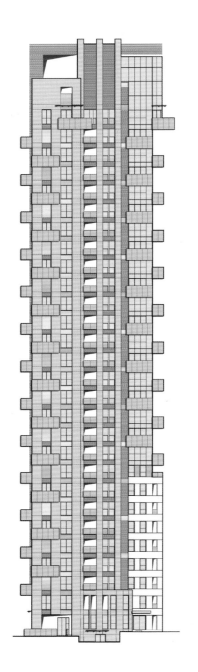
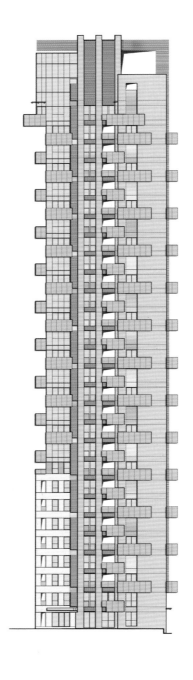

Elevation
立面图

Tower 1 10~27 Floor Plan(Odd Floors)
1栋 10~27 层奇数层平面图

Section
剖面图

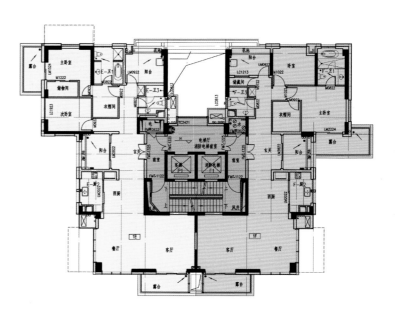

Tower 1 10~27 Floor Plan(Even Floors)
1栋 10~27 层偶数层平面图

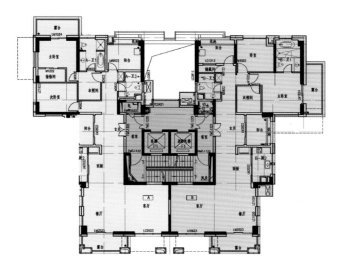

Tower 1 4~8 Floor Plan(Odd Floors)
1栋 4~8 层奇数层平面图

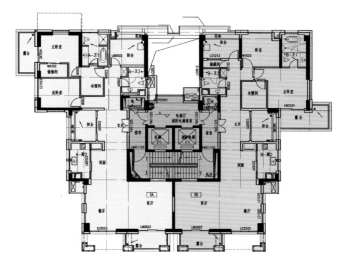

Tower 1 4~8 Floor Plan(Even Floors)
1栋 4~8 层偶数层平面图

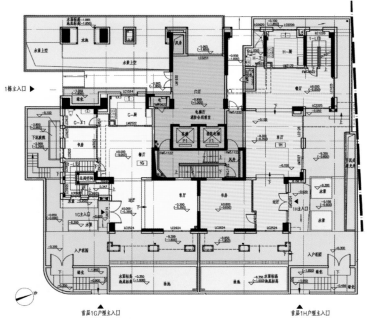

Ground Floor Plan
首层平面图

2nd Floor Plan
2层平面图

SHAPE 造型

STRUCTURE 结构

SUSTAINABILITY 可持续性

FAÇADE 立面

WKL Hotel and Residences, Kuala Lumpur, Malaysia

马来西亚吉隆坡 WKL 酒店与住宅大厦

Architect: SOM
Location: Kuala Lumpur, Malaysia
Floors: 55
Renderings: SOM, Crystal CG

设计公司：SOM 建筑设计事务所
地点：马来西亚吉隆坡市
层数：55
效果图：SOM 建筑设计事务所、Crystal CG

This 55-storey tower will be the home of the WKL Hotel in Kuala Lumpur. Its presence emulates W's brand motto: "inspiring, iconic, innovative and infuential."

这栋高 55 层的建筑将会成为 WKL 酒店在吉隆坡的新的标志性建筑。它的存在传达了 WKL 品牌的座右铭："激励、形象、创新和影响力"。

The building is located on Jalan Ampang, a major downtown thoroughfare, in very close vicinity to the KL Civic Centre and the Petronas Towers. Convenient traffic can not only make the architecture more popular, but bring good returns for the hotel. At the same time, the height of the hotel gives it a broad perspective. Standing at the upper level, people will have a panoramic view of beautiful scenery from the city. That people can enjoy the landscape when they are having a rest is a dual purpose.

A five-star hotel, WKL boutique hotel includes 150 rooms and "WOW" suites, a ballroom, meeting rooms, destination lounges and signature restaurants. A residential tower above includes 353 units and residential amenities. Indoor luxury decoration can let people feel modern city life anytime and anywhere. Room size varies from large to small, and visitors are free to choose what ever environment they want to live in. At the same time, the configuration of leisure and recreational facilities such as swimming pool are the perfect embellishment for the building.

The building program is distributed vertically in a 12-storey podium and a 43-storey tower. The podium includes seven garage levels plus the hotel ballroom facilities. The hotel destination lounge, spa and outdoor pool are located on the podium rooftop. The tower includes 10 storeys of hotel guestrooms and 28 storeys of residential units. The top floor level is dedicated to the residential amenities and an outdoor pool. The tower's square floor plate offers 360-degree views of the surrounding cityscape.

The WKL Hotel and Residences projects a distinctively dynamic, contemporary image, in tune with its brand and the transformation of Kuala Lumpur centre into a world destination.

本项目坐落于马来西亚吉隆坡市中心的交通要道——安邦路，与文娱中心和双子塔隔街相望。便利的交通不仅可以为建筑打响知名度，更为酒店带来了良好的收益。同时，酒店的高度让其拥有了广阔的视野。站在酒店上层，将来自城市的绝美风光尽收眼底，让人们在休息的时候可以欣赏到外部的景观。

WKL 酒店是一座五星级酒店，拥有 150 套客房和总统套房、宴会厅、会议室、休息室和餐厅。住宅楼部分有 353 套客房，并配置了住宅设施。室内奢华的装饰，让人们可以随时随地感受到来自现代的都市生活。房间从大到小，供自由挑选，旅客可以自由地选择想要居住的环境。同时，游泳池等休闲娱乐设施的配置，更是对建筑的绝佳点缀。

建筑分为两部分：12 层的裙楼和 43 层的主楼。裙楼有 7 层的停车场，也同时具有酒店宴会厅的功能。休息室、水疗和户外泳池都位于裙楼的屋顶。建筑共有 10 层的客房和 28 层的住宅单位。顶层设有居家福利设施和户外泳池。方形的楼面板为建筑提供了 360 度的城市视野。

WKL 酒店和住宅项目拥有独特的动态现代设计，符合其品牌和将吉隆坡中心转换成为世界目的地的理念。

FEATURE 特点分析

STRUCTURE

The project is located in the centre of Kuala Lumpur City, and how to arrange each functional area of the hotel and service area in the single building will be a big challenge. Podium is the parking and Banquet Hall, and on the roof there is a lounge, Spa and outdoor swimming pool. The middle of the building is for guest rooms. Excellent design provides more comprehensive services for their customers.

结构

项目在吉隆坡市中心，在单体的建筑中如何分配酒店的各个功能区与服务区将成为挑战。裙楼是停车场和宴会厅，屋顶有休息室、水疗和户外泳池等。主楼中部为客房。出色的设计为顾客提供更全面的服务。

FAÇADE 立面

SHAPE 造型

MATERIAL 材料

ENERGY-SAVING 节能

HQ by Sansiri, Bangkok, Thailand

泰国曼谷 HQ by Sansiri 住宅大厦

Architect: dwp | design worldwide partnership
Client: Sansiri PLC
Location: Bangkok, Thailand
Gross Floor Area: 24,240 m²
Floors: 36
Realization: 2014

设计公司：世界级设计公司 dwp
客户：Sansiri PLC
地点：泰国曼谷市
建筑面积：24 240 平方米
层数：36
完成时间：2014 年

World-class architecture and interior design firm dwp created the luxury architecture and interiors for the entire HQ by Sansiri development in the Thai capital. This high-end residential tower stands in the popular Thong Lo district of the Bangkok city centre at 36 floors high, designed under the theme that "Sophistication is the Ultimate Form of Luxury".

世界级的建筑及室内设计公司 dwp 在泰国首都为 HQ by Sansiri 的全部项目创造了奢华的建筑和室内设计。这座极品的住宅塔楼坐落在曼谷市中心的 Thong Lo 区，楼高 36 层，设计主旨是"混搭锻造极致奢华"。

Master Plan
总平面图

In addition to the highly identifiable nature of the building exterior, the façade is also defined by a feature angle beam, one of the client's favourites. The lean structure highlights natural materials, and is surrounded by water bodies, as well as the large ground floor garden. As with an artwork located on the ground of a museum, HQ by Sansiri becomes the art sculpture of this garden environment. The overall impression is of an understated, yet knowingly, chic, stylish ambience.

Compact and functional, the building boasts modern condos and is geared for both functionality and aesthetics, in all aspects and areas. Inside and out the building reflects both superiority and subtlety. The swimming pool is a unique and intriguing addition, with infinity style and daybed lounges incorporated, to feel not only that you are floating on the water, but hovering on the skyline. With added greenery, residents can continuously feel like one with nature. One further outstanding feature of the pool, is that in the middle, a glass box seemingly emerges. This houses the communal fitness studio, complete with high-tech equipment. For adding convenience for residents in this bustling city, ample separate parking has been provisioned. Within the lobby area, the focus is on the extensive use of imported marble and other natural stones, on flooring and a stone feature wall.

There is an assured connectivity between internal and external spaces, with features and décor to hone the experience. Architectural detailing is not limited to just the façade. Meticulous space planning involves a symmetrical axis and clean elevation, while highlighting easy accessibility. The tower was furthermore conceived for good ventilation and maximum natural light throughout, to allow for utmost energy-saving and environmental considerations. All rooms in all apartments have maximised windows, which is in keeping with the minimalist theme, and also affords optimal views of the city and garden scape. This building is the first of its kind in the area, with an emphasis on providing a natural environment for residents. The design concept has seen a marked success, with the project being sold out within merely 2 days after the sales launch.

FEATURE 特点分析

SHAPE

Inspiration for the design concept is drawn from the idea of minimalism and the pioneering architectural style of Ludwig Mies van der Rohe. The tower erupts as a sharp blade, piercing the horizon, standing proudly and uniquely within the cityscape. As a stunning art sculpture, the building structure makes an understated, yet impressive impact, emphasizing a sense of nature given back to the community.

造型

设计理念的灵感来自于极简派艺术概念和路德维格的先锋派建筑风格。塔楼宛若锋利的刀片，划破地平线，迸发出来，独特而骄傲地屹立在城市的风景之中。作为一件特有的艺术雕刻品，这座建筑结构不仅给人以低调的印象，却也加强了社区回归自然的感觉。

Ground Floor Plan
首层平面图

Lobby Section
大堂剖面图

建筑不仅拥有独特的外观，其正立面也是由特定的角钢梁所定义，这正是客户的最爱之处。塔楼结构精细，用料顶级天然，周围水体环绕，还有超大的地面花园，就像一件艺术品坐落在博物馆的地面上。HQ by Sansiri 成了这处花园环境的艺术雕塑。塔楼给人低调的整体印象，却不失智慧、别致、时髦的气氛。

该建筑是座现代公寓，每个方面都兼具功能和美学性能。建筑内外处处都反映出它的优质精细。游泳池很独特，里面有躺椅和休息处，让人感觉不但飘漂在水上，还悬浮在空中。处在更多绿色植物中的居民能时常体验与大自然融和一体的感觉。游泳池更显著的特点是位于池中间的玻璃盒。这是个公共健身室，配有完整的高科技设备。为了给生活在忙乱的城市里的居民进一步提供便利，这里提供了足够车位的独立停车场。在大堂里，地板和有特色的石墙广泛使用进口大理石和其他天然石材，尤其引人注目。

室内和户外空间有着密切的联系，各种特色和装饰能给人们带来丰富的体验密切的联系。建筑细节不仅局限于立面的设计，还包括对建筑对称轴和立面高度的空间规划，并突出其可及性。此外，塔楼对通风、最大程度的自然光照方面有很好的构思，以利于最大程度的节能环保。公寓所有的房间窗户都足够大，以保持极简抽象派艺术建筑的主题，同时又对城市和花园空间获取最佳视角。这座建筑是这一区域第一座强调为居民提供自然居住环境的高楼。设计理念取得了显著的成功，项目开始销售仅仅两天就已全部售完。

2 BEDROOM TYPE B
SEE NO. A-4516

1 BEDROOM TYPE D
SEE NO. A-4505

1 BEDROOM TYPE I_UPPER
SEE NO. A-4514

10th Floor Plan
10 层平面图

SHAPE 造型

STRUCTURE 结构

ROOFTOP GAEDEN 屋顶花园

FAÇADE 立面

M Ladprao, Bangkok, Thailand

泰国曼谷拉普绕酒店

Architect: P&T Group
Client: Major Development PCL
Location: Bangkok, Thailand
Site Area: 3,148 m²
Gross Floor Area: 25,133 m²
Floors: 42
Realization: 2013

设计公司：P&T 集团
客户： Major Development PCL
地点：泰国曼谷市
占地面积：3 148 平方米
建筑面积：25 133 平方米
层数：42
完成时间：2013 年

M Ladprao elegantly rises 42 floors towards the sky of Bangkok, embracing dynamic views of Chatuchak Park and its urban surrounds. A dynamic landmark derived from the transformation of cubic interlocking planes, the development comprises of a high-rise residential beacon with 2-storey low-rise separate retail premises.

高 42 层的 M Ladprao 建筑优雅地向曼谷的高空延伸，直通天际，环抱来自恰图恰克公园和城市的美景。环环相扣的立体面板造就了建筑的地标性标志。项目共包括高层住宅和 2 层高的低层独立商场。

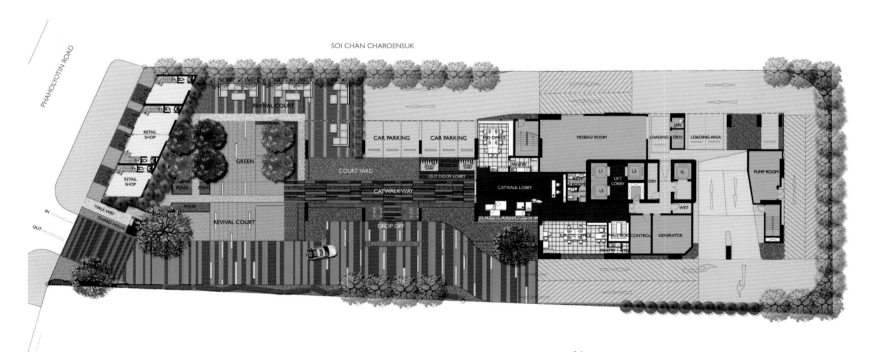

SOI CHAN CHAROENSUK

PHAHOLYOTIN ROAD

RETAIL SHOP

RETAIL SHOP

REVIVAL COURT

GREEN

COURT YARD

CATWALK WAY

REVIVAL COURT

DROP OFF

POND

POND

WALK WAY

GUARD HOUSE

IN

OUT

CAR PARKING

CAR PARKING

HD OFFICE

OUT DOOR LOBBY

PANTRY

CATWALK LOBBY

LIFT LOBBY

L1   L3

L2

FL

DRY

WET

MAIL BOX

CONTROL

GENERATOR

PANTRY

ESTIC OFFICE

MOB/HV ROOM

JAN

LOADING DRIV

LOADING AREA

PUMP ROOM

N

Ground Floor Plan
首层平面图

0 1   5   10

The mixed-use development embraces the modern luxury of living, while encapsulating the advanced social and environmental values of the modern Bangkok urbanite. Its residents are invited to quench their social appetite at a dynamic roof top sky bar while a cantilevered swimming pool, club facilities such as fitness, running track and library as well as ample green spaces are located on a lower podium roof deck.

With 9 residential units per floor, M Ladprao features 289 one or two bedroom units with ample parking provision located on the fifth floor within a dedicated and inviting podium area. Located within walking distance from Central Plaza Ladprao and Tesco Lotus, amenities are within easy reach.

Large open green areas at ground floor separate the main residential tower from the low-rise retail unit along Phahonyothin road, while providing its users with outdoor spaces for socializing. The 2-storey retail unit sits proud as a separate entity at the main entrance of the development and mimics the development's outdoor living language by providing terraced al fresco dining spaces.

多功能的综合建筑符合现代奢华生活的理念，同时包含了现代曼谷都市人先进的社会和环境价值。居民可以在动感的屋顶天空吧娱乐、社交；同时，悬臂游泳池、健身俱乐部等设施，如跑道和图书馆，以及绿色葱郁空间则位于较低的屋顶甲板上。

M Ladprao 每层有 9 个住宅单元，特色在于拥有 289 个一室或两室的住宅单元，并在第五层配有充足且专用的停车位。距离拉普绕中央广场和莲花超市步行就可前往，且到达各便利设施也方便快捷。

一个庞大的开阔绿色区域沿着拍凤裕庭路将主要住宅建筑楼和底层零售单元分开来，同时为它的使用者提供了户外社交场所。这个两层的零售单元作为一个副入口，位于项目的主要入口处，并通过提供露台用餐空间，形象地表现出建筑的户外生活语言。

FEATURE 特点分析

STRUCTURE

The architectural design gives people unlimited longings to enjoy the most beautiful scenery around the city super on this building high up to the sky. Luxury decoration in line with the concept of modern life allows people to experience that kind of fashion and elegance. In addition, full facilities can meet the maximum demands of people.

立面

建筑设计给人无限的渴望去享受城市中高达天际的大厦上方最美丽的风景。奢华的装饰与现代生活理念相符合，使人们经历那种时尚和优雅。另外，全面的设施能满足人们需要的最大化。

Alternative Typical Plan
典型楼层平面图

3rd Typical Floor Plan
3 层典型楼层平面图

ECOLOGY 生态

FAÇADE 立面

SUSTAINABILITY 可持续性

SHAPE 造型

The Pinnacle @ Duxton, Singapore

新加坡达士敦坪高楼组屋

Architect: RSP Architects Planners & Engineers (Pte) Ltd, Arc Studio Architecture + Urbanism
Location: Singapore
Site Area: 25,172.1 m²
Gross Floor Area: 235,597 m²
Floors: 50
Height: 172 m
Photography: RSP Architects Planners & Engineers (Pte) Ltd, Arc Studio Architecture + Urbanism
Drawings: Arc Studio Architecture + Urbanism

设计公司：雅思柏设计事务所、Arc Studio 建筑与都市设计事务所
地点：新加坡
占地面积：25 172.1 平方米
建筑面积：235 597 平方米
层数：50
高度：172 米
摄影：雅思柏设计事务所、Arc Studio 建筑与都市设计事务所
图纸：Arc Studio 建筑与都市设计事务所

Soaring at 50 storeys, Pinnacle @ Duxton redefines high-rise high-density living and challenges the conventions of public housing as an architectural typology. The project addresses pragmatic, financial, social issues, and responds sensitively to a myriad of planning constraints. It boldly demonstrates a sustainable and liveable urban high-rise high-density living and initiates an innovative typology of public communal spaces that are metaphorically reclaimed from the air.

这座 50 层楼高的住宅，重新诠释了超高层、高密度生活方式。作为一类建筑的象征，同时打破了公共组屋的常规建筑方式。该项目提出实用、经济、社会层面的课题，对复杂的规划限制条件巧妙地给予回应。大胆地体现了可持续和宜居都市的高层、高密度的生活方式，创造了一种空中共享空间的新形式——也就是巧妙地利用天空。

FAÇADE

Efficiently constructed off-site, pre-fabricated concrete building components are delivered and put together on the tight site. Residents are given an unprecedented choice of exterior façade treatments—planter boxes, bays, bay windows, windows, and balconies. By turning the façade into a system of panels incorporating structure and services, and with a simple and affordable application of paint finish, a highly differentiated façade is created from an undifferentiated fabrication process —creating visual delight and reducing the perceived building mass.

立面

预制混凝土建筑部件在场外制造后再运送过来，根据场地的紧凑地形进行组装。设计师以别样的处理手法为居民带来了耳目一新的立面。立面上包含了花盆、凸窗、窗口和阳台。通过将立面转化成一个结构与服务相结合的面板系统，再配合简单又实惠的漆面和分化制造工艺的应用，让建筑的的立面真正独树一帜，既带来愉悦的视觉感，又减少了建筑的沉重感。

When looked at from flank, building structure is like three huge picture frames with continuous arrangement, stands in the urban's greening network, cases distant blue sky, at the foot of the building among green trees, revealing the several red tops of houses. These are accessory buildings at the high-rise positive square that are underground garage and equipment room, and the height of the buildings is the same as the first layer level of main body . Accessory buildings are designed into slope-form top and decorated into fluctuating vestibular garden. Between roads leading to each unit paving the large turf for residents to take a break, entrance roads of units are paved with floor tile. Both sides on the grass are regularly planted with high broadleaf trees , echoing with sinking entry in height. Each unit is linked with an open corridor. On wall body side of corridor, stonecrop plants and leafy herbaceous plants are chosen. Sandwiched between a few plants are half tall trees, strewing at random , neat and pure and fresh. The building and environment are fused into one local landscape.

从侧面望去，建筑结构像三个连续排列的巨大镜框，耸立在城市绿化的树丛中，框住了远处的蓝天。建筑物脚下，绿树中散露着几处红顶的保留店屋。高楼正面的梯形广场上，是地下车库、设备房等附属建筑，这些建筑的高度与主体的首层平齐。附属建筑设计成坡形斜顶，布置成起伏有致的前庭花园，在各单元道路之间铺设了大片草皮，供居民憩息，单元入口道路用地砖铺面，两旁草地上有规律地种植了较高的阔叶乔木树种，在高度上与下沉的入口形成呼应。各单元楼之间以敞开式的内连廊相通，连廊靠墙体一侧选用景天类、观叶类草本植物，间夹着几株半高的小树，错落有致，整洁清新，楼体与环境自然融合成一体。

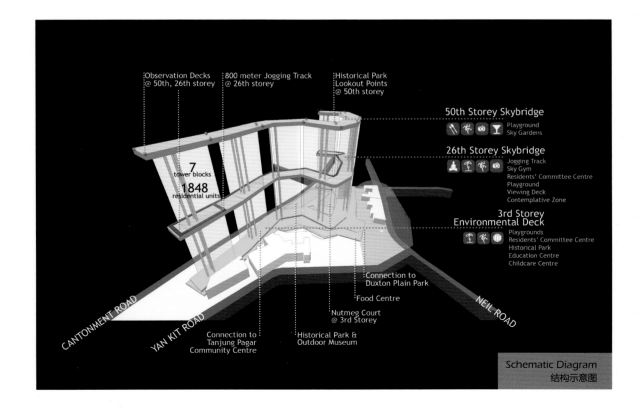

Observation Decks
@ 50th, 26th storey

800 meter Jogging Track
@ 26th storey

Historical Park
Lookout Points
@ 50th storey

50th Storey Skybridge
Playground
Sky Gardens

26th Storey Skybridge
Jogging Track
Sky Gym
Residents' Committee Centre
Playground
Viewing Deck
Contemplative Zone

3rd Storey
Environmental Deck
Playgrounds
Residents' Committee Centre
Historical Park
Education Centre
Childcare Centre

7
tower blocks

1848
residential units

Connection to
Duxton Plain Park

Food Centre

Nutmeg Court
@ 3rd Storey

Historical Park &
Outdoor Museum

CANTONMENT ROAD

YAN KIT ROAD

Connection to
Tanjung Pagar
Community Centre

NEIL ROAD

Schematic Diagram
结构示意图

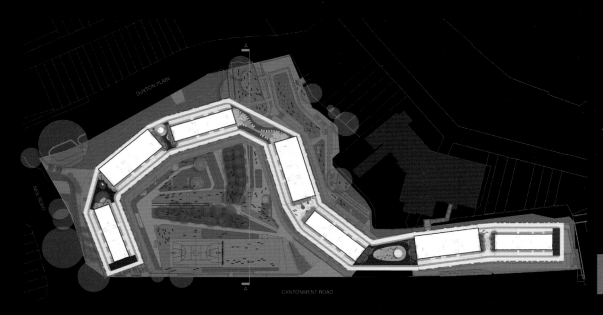

Site Plan
总平面图

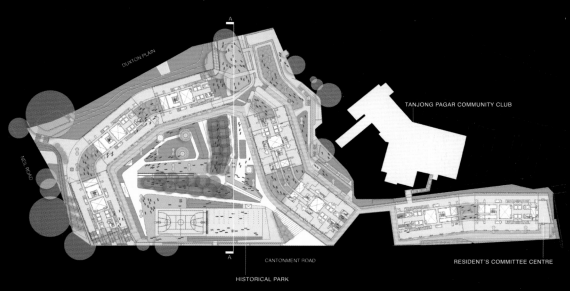

TANJONG PAGAR COMMUNITY CLUB

HISTORICAL PARK

RESIDENT'S COMMITTEE CENTRE

3rd Storey Plan
3 层平面图

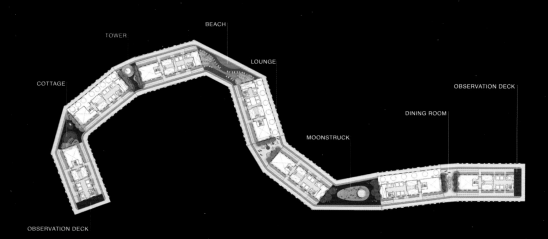

BEACH

TOWER

LOUNGE

COTTAGE

OBSERVATION DECK

DINING ROOM

MOONSTRUCK

OBSERVATION DECK

26th Storey Plan
26 层平面图

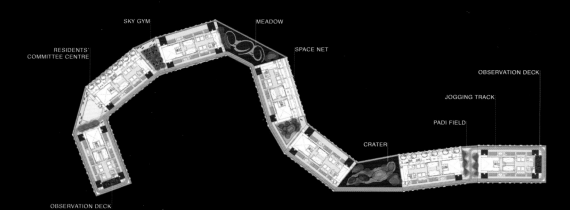

SKY GYM

MEADOW

RESIDENTS'
COMMITTEE CENTRE

SPACE NET

OBSERVATION DECK

JOGGING TRACK

PADI FIELD

CRATER

OBSERVATION DECK

50th Storey Plan
50 层平面图

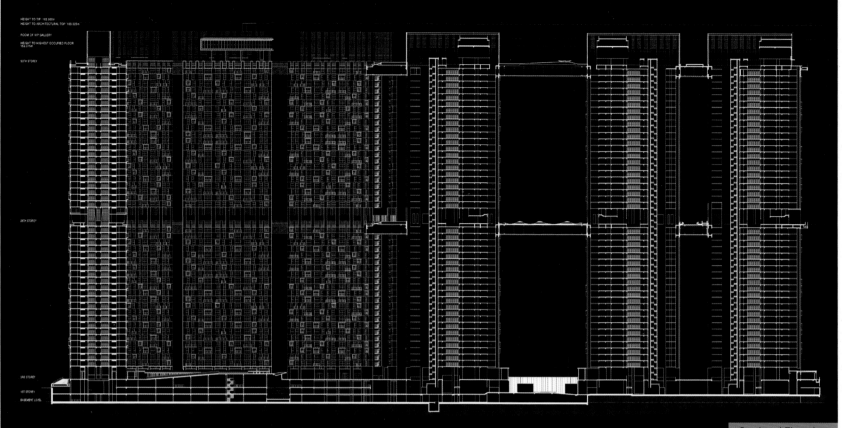

HEIGHT TO TIP 163.500m
HEIGHT TO ARCHITECTURAL TOP 163.325m

ROOM OF VIP GALLERY

HEIGHT TO HIGHEST OCCUPIED FLOOR
153.375m

50TH STOREY

26TH STOREY

2ND STOREY
1ST STOREY
BASEMENT LEVEL

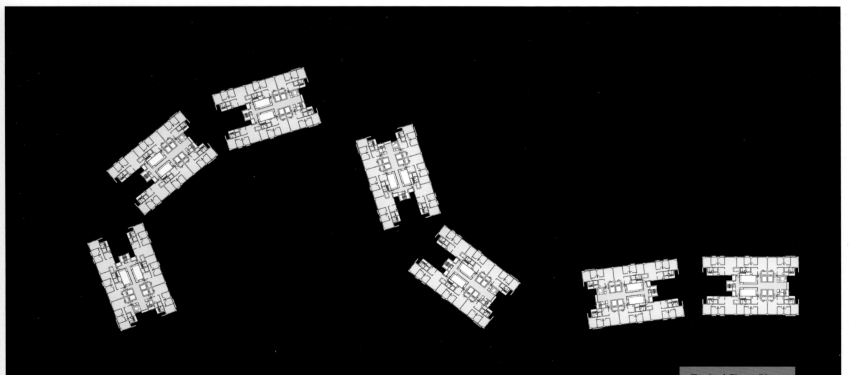

(a.01)
Plug in: Bay Window
A ledge extends 3/4 height bay window outwards from the facade. Decorate it with your own private collection of

(b.01)
Maximise openings at living rooms hence oscillation of window panels only occur at the bedrooms

(c.01)
Plugin
Distribution: Bay
Window

(c.02)
Plugin
Distribution: Bay

(a.02)
Plug in: Bay
Enjoy expansive views with the floor to ceiling window extending from the inside of the apartment.

(b.02)
Taking alternate sets from rows A & B (Aa Bb Ac). Taking alternate sets from rows A & B. Maximising openings (Aa Bb Bc)

(a.03)
Plug in: Balcony
Step into the balcony with a glass of wine in hand & watch the world go by many storeys below.

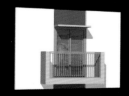

(b.03)
Let 0 = Base Position. 1 = Shift Right.

(c.03)
Plugin Distribution:
Balcony

(c.04)
Plugin
Distribution:
Planter

(a.04)
Plug in: Planter
Nurture your favourite plants & watch them grow merrily in the planter just outside the window.

(b.04)
Inserting balconies approx 5.9m per row. Not more than 1 Balcony per column.

1 x unit   1 x unit   1 x unit

Facade Diagram
立面结构图

The new hill is a lush environmental deck that connects strategically with the existing urban network while forming a green lung for the city. Layers of tree screens border the site and pathways to provide varying degrees of opacity and privacy, softening the massiveness of the towers. On the 26th and 50th storeys, continuous Sky Gardens weave through seven tower blocks, forming a simple yet powerful sculptural skyline. These thematic gardens are an extension of the living environment for residents and provide diverse spaces for community interaction while overlooking the city. They also function as areas of refuge during fires and allow the sustainable sharing of mechanical services, reducing seven sets to just three.

Along the outside of the terrace of the 26th layer is fixed with white iron guardrail. There is natural batten floor along inner of terrace. The layer between the wall and terrace is covered with turf. When you stand on natural batten with green grass behind and take an overlooking on the city landscape, immediately it lets a person relax fully, relaxing in the bosom of nature.

青翠繁茂的植物环境小丘与现有的城市化绿网相连，形成了一处新的城市绿肺。树荫覆盖的场所和小路提供了幽深隐密的去处，亦使深沉的高楼变得相对柔和。在第 26 和第 50 层楼层，连续的空中花园穿过七座楼体，形成了雕刻般简洁有力的天际线。这些主题花园俯视着城市，是居民生活环境的延伸，也为社区互动提供了多样空间，发生火灾时，它们又是逃生避难场所。花园还使空调机械得以持续共享，从七部减少到三部。

26 层的露台外沿竖立着白色铁制护栏，内沿向里采用自然木条铺地。木条地后连接着墙体的地表覆盖了草皮，背依绿色草皮。当你脚踏天然木条，凭栏远眺城市景观，顿时让人松驰了全部神情，轻松投入了大自然的怀抱。

SHAPE 造型

FAÇADE 立面

BALCONY 阳台

SKY GARDEN 空中花园

V on Shenton, Singapore

新加坡珊顿大道五号

Architect: *UNStudio*
Client: *UIC Investments (Properties) Pte Ltd*
Location: *Singapore*
Site Area: *6,778 m²*
Gross Floor Area: *85,507 m²*
Floors: *53 (residential tower), 23 (office tower)*
Height: *237 m (residential tower), 123 m (office tower)*

设计公司：UNStudio
客户：UIC 私人投资有限公司
地点：新加坡
占地面积：6 778 平方米
建筑面积：85 507 平方米
层数：53（住宅楼），23（办公楼）
高度：237 米（住宅楼），123 米（办公楼）

The former UIC building has dominated the city skyline as Singapore's tallest building for many years since its completion in 1973 and is part of an important collection of towers located along Shenton Way in the heart of Singapore's Central Business District. Today, the area is undergoing rejuvenation and transformation and "V on Shenton", the new UIC building, forms a part of this redevelopment. The dual programming of "V on Shenton", comprising office and residential, presents a unique situation in this area of the city.

著名的 UIC 大厦建成于 1973 年，多年来一直作为新加坡最高的建筑占据着城市的天际线，它也是坐落在新加坡商业中心区核心地段的珊顿大道两旁重要的塔楼群中的一部分。今天这一区域正在发生改变并焕发着新活力：珊顿五号——一座新的 UIC 大厦，就是再开发计划的一部分。珊顿五号的双重发展规划包含办公和住宅，在城市的这一区域显示了其独特的优势。

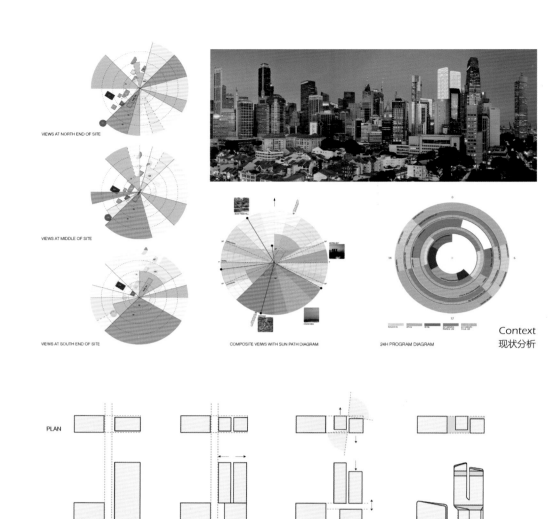

VIEWS AT NORTH END OF SITE

VIEWS AT MIDDLE OF SITE

VIEWS AT SOUTH END OF SITE

COMPOSITE VIEWS WITH SUN PATH DIAGRAM

24H PROGRAM DIAGRAM

Context
现状分析

PLAN

ELEVATION

massing development
体量分析

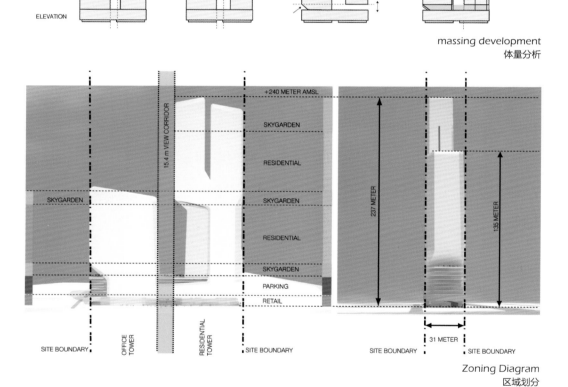

+240 METER AMSL

15.4 m VIEW CORRIDOR

SKYGARDEN

RESIDENTIAL

SKYGARDEN

RESIDENTIAL

SKYGARDEN

PARKING

RETAIL

OFFICE TOWER

RESIDENTIAL TOWER

237 METER

135 METER

31 METER

SITE BOUNDARY

SITE BOUNDARY

SITE BOUNDARY

SITE BOUNDARY

Zoning Diagram
区域划分

The twin tower of "V on Shenton" is comprised of a 23-storey office building and a 53-storey residential tower, with the dual programming of the building highlighted through its massing. The office tower corresponds to the scale of the surrounding buildings the streets. The angle in the office tower roof corresponds to the upper of two mid-level sky gardens of the residential tower.

Just as the office and residential towers are of the same family of forms, so do their façades originate from the same family of patterns. These geometric panels add texture and cohesion to the building, while reflecting light and pocketing shade. The texture and volume of the façade are important for maintaining the comfort of those living and working in the residential and office buildings. Shading devices and high-performance glass are important for developing a sustainable and liveable façade.

Each tower is framed by "chamfers": a line that unifies the composition of the residential tower, the office tower and the plinth. During daytime, the chamfer appears smooth in contrast to the textured surfaces of the towers. At night, the chamfer lights up as a continuous line framing the towers, car park and sky gardens. The chamfers at the north end of the office tower also open up the corners to views of Marina Bay, Buket Timah Hill and the Central Business District.

The building rises up to distinguish itself from the neighbouring buildings. Above this sky lobby the unit mix of the residential tower changes with a subtle display of its split core. This separation also exhibits the natural ventilation concept of the tower which is further effected through ventilation slots next to the cores. Distinctive to "V on Shenton", these slots are covered by façade cladding with openings for air circulation, resulting in a continuous, uninterrupted hexagonal façade pattern on the residential tower.

珊顿五号的双塔包含一座 23 层的办公塔楼和一座 53 层的住宅塔楼，建筑的双重规划通过其体量突出显现出来。办公楼与它周围建筑和街道的规模相呼应，办公塔楼顶部的角度是为回应住宅塔楼中部的两个空中花园。

正如办公和住宅塔楼有着相同的造型，它们的立面也有着同样的图案。这些几何形的面板，使得建筑质地轻巧内敛，外形光亮袖珍，建筑表面的纹理和外形对维持住宅和办公楼内生活和工作的舒适起到非常重要的作用。遮荫装置和高性能的玻璃对可持续的宜居住宅外表立面也相当有用。

每座塔楼都有一定程度的倾斜，将住宅塔楼基座和办公塔楼基座连成一体。白天对应着网状表面的楼体，楼顶是平顺的斜面。晚上，斜面顶灯点亮，像一条连续的椎线勾勒出塔楼、停车场和空中花园。办公塔楼北边的斜楼顶也开放，观看朝向滨海湾的金沙酒店、希克蒂玛大厦和中央商务区。

本项目耸立的塔楼在周围建筑中独树一帜。住宅塔楼在空中大堂以上的部位发生变化，精确地分成两座。这样的分离是为了便于塔楼的自然通风，使紧邻中心的通风孔更好地发挥作用。珊顿大道五号的与众不同在于这些通风孔被建筑表皮所覆盖，开口加强使空气循环，并在住宅塔楼的表面形成了连续的六边形图案。

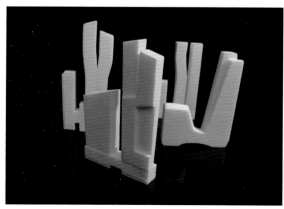

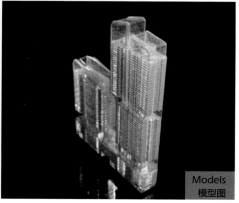

Models
模型图

PROGRAM DISTRIBUTION

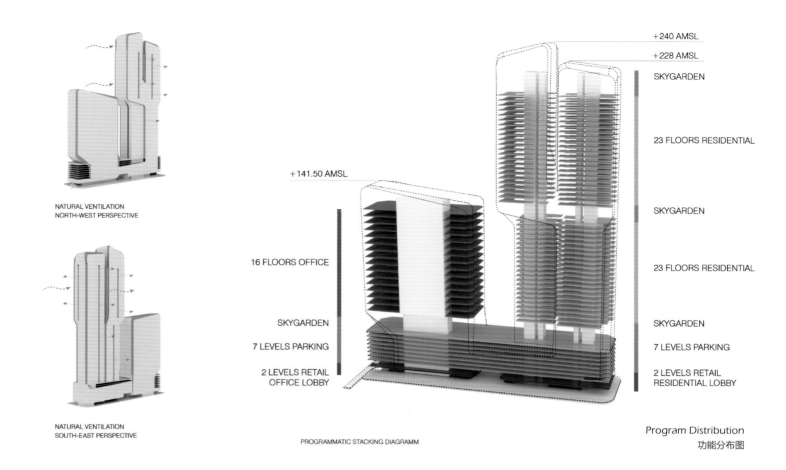

NATURAL VENTILATION
NORTH-WEST PERSPECTIVE

NATURAL VENTILATION
SOUTH-EAST PERSPECTIVE

PROGRAMMATIC STACKING DIAGRAMM

+ 141.50 AMSL

16 FLOORS OFFICE

SKYGARDEN

7 LEVELS PARKING

2 LEVELS RETAIL
OFFICE LOBBY

+ 240 AMSL

+ 228 AMSL

SKYGARDEN

23 FLOORS RESIDENTIAL

SKYGARDEN

23 FLOORS RESIDENTIAL

SKYGARDEN

7 LEVELS PARKING

2 LEVELS RETAIL
RESIDENTIAL LOBBY

Program Distribution
功能分布图

BUILDING ENVELOPE CONCEPT

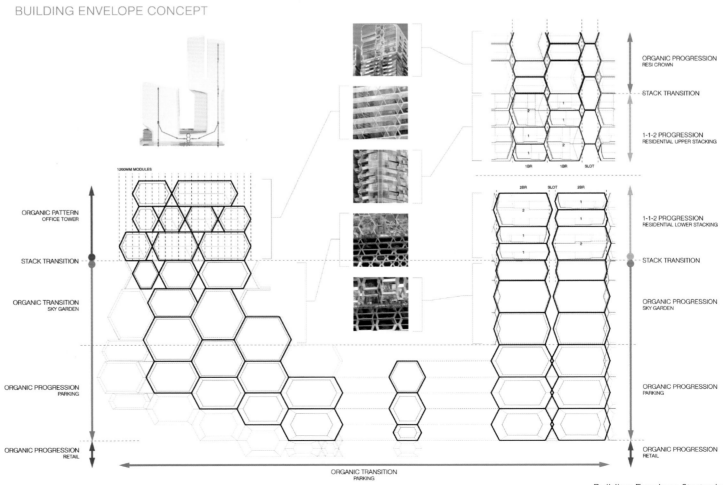

1200MM MODULES

ORGANIC PATTERN
OFFICE TOWER

STACK TRANSITION

ORGANIC TRANSITION
SKY GARDEN

ORGANIC PROGRESSION
PARKING

ORGANIC PROGRESSION
RETAIL

ORGANIC TRANSITION
PARKING

ORGANIC PROGRESSION
RESI CROWN

STACK TRANSITION

1-1-2 PROGRESSION
RESIDENTIAL UPPER STACKING

1BR   1BR   SLOT

2BR   SLOT   2BR

1-1-2 PROGRESSION
RESIDENTIAL LOWER STACKING

STACK TRANSITION

ORGANIC PROGRESSION
SKY GARDEN

ORGANIC PROGRESSION
PARKING

ORGANIC PROGRESSION
RETAIL

Building Envelope Strategies
建筑立面结构概念

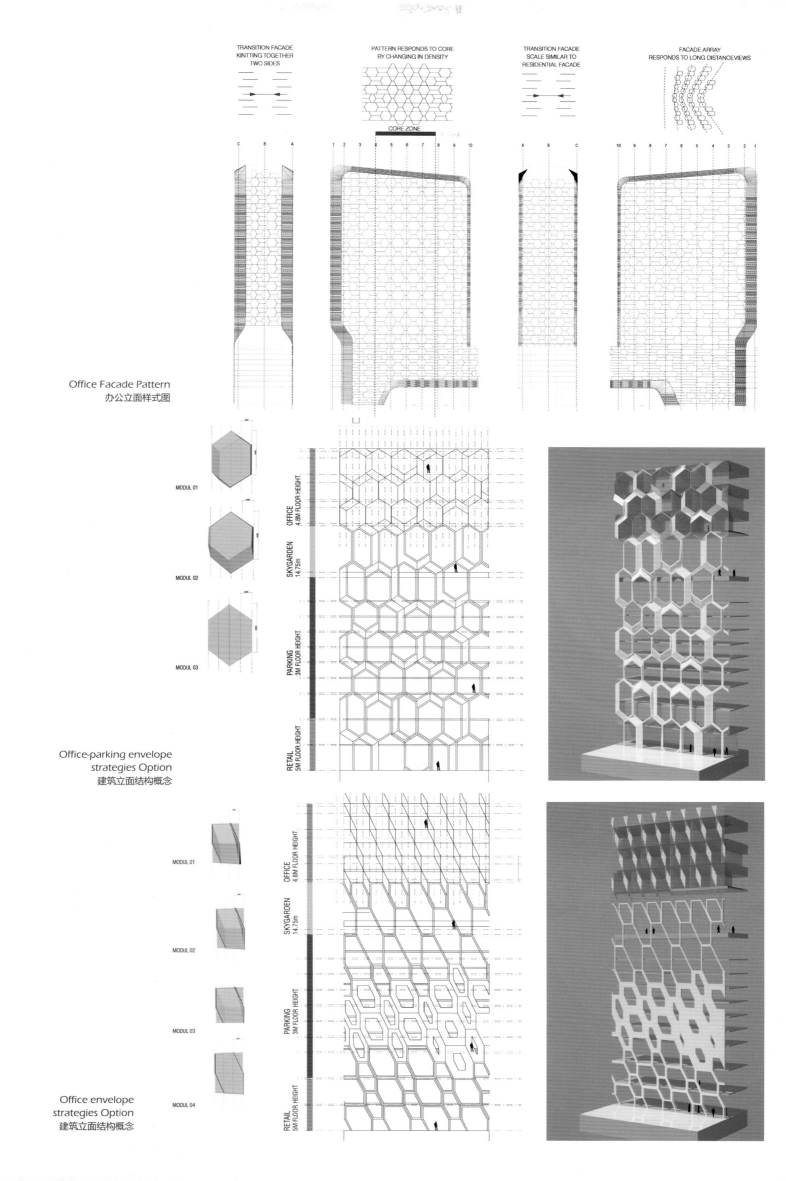

TRANSITION FACADE
KINTTING TOGETHER
TWO SIDES

PATTERN RESPONDS TO CORE
BY CHANGING IN DENSITY

CORE ZONE

TRANSITION FACADE
SCALE SIMILAR TO
RESIDENTIAL FACADE

FACADE ARRAY
RESPONDS TO LONG DISTANCEVIEWS

Office Facade Pattern
办公立面样式图

MODUL 01

MODUL 02

MODUL 03

OFFICE
4.8M FLOOR HEIGHT

SKYGARDEN
14.75m

PARKING
3M FLOOR HEIGHT

RETAIL
5M FLOOR HEIGHT

Office-parking envelope
strategies Option
建筑立面结构概念

MODUL 01

MODUL 02

MODUL 03

MODUL 04

OFFICE
4.8M FLOOR HEIGHT.

SKYGARDEN
14.75m

PARKING
3M FLOOR HEIGHT

RETAIL
5M FLOOR HEIGHT

Office envelope
strategies Option
建筑立面结构概念

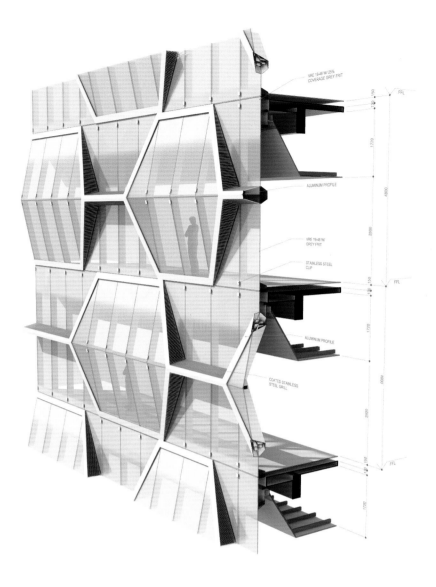

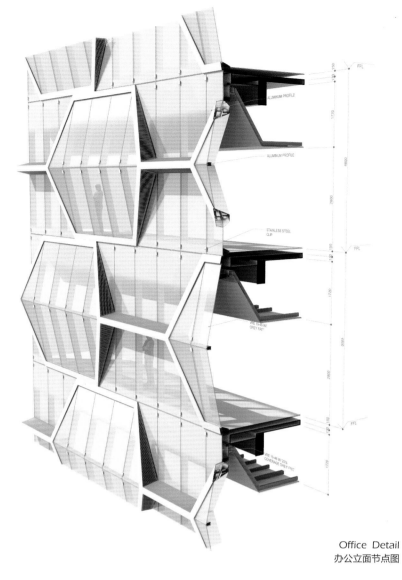

Office Detail
办公立面节点图

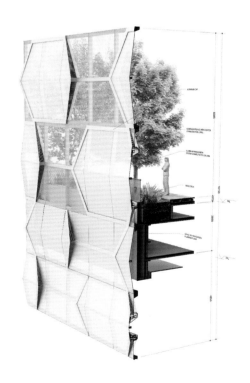

Office SG Detail
办公立面细节效果图

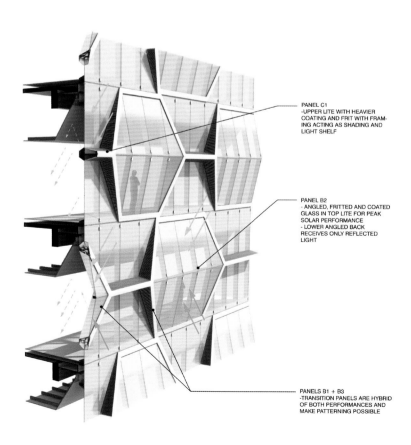

PANEL C1
-UPPER LITE WITH HEAVIER COATING AND FRIT WITH FRAMING ACTING AS SHADING AND LIGHT SHELF

PANEL B2
- ANGLED, FRITTED AND COATED GLASS IN TOP LITE FOR PEAK SOLAR PERFORMANCE
- LOWER ANGLED BACK RECEIVES ONLY REFLECTED LIGHT

PANELS B1 + B3
-TRANSITION PANELS ARE HYBRID OF BOTH PERFORMANCES AND MAKE PATTERNING POSSIBLE

Office Facade Environmental 3D
办公立面环境 3D 图

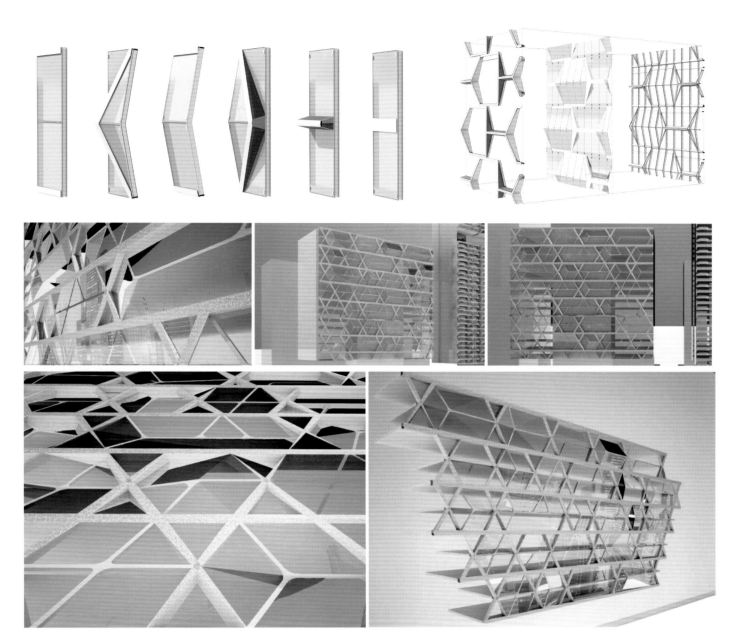

FEATURE 特点分析

FAÇADE

The basic shape of the hexagon is used to create patterns that increase the performance of the façades with angles and shading devices that are responsive to the climatic conditions of Singapore. The office tower is based on a curtain wall module and an optimised number of panel types, recombined to create a signature pattern. In contrast, the residential façade is based on the stacks of unit types. The pattern of the residential façade is created by the incorporation of the residential programme (balcony, bay window, planter and a/c ledge) and the combination of one or two storey high modules with systematic material variations.

立面

建筑基本立面采用六边形造型以增强立面的遮荫角度和功能，以适应新加坡的气候条件。办公塔楼的立面是一面幕墙，由最佳数量的显示屏类型的平板构成，以形成其特征。与此相对比，住宅塔楼的立面基于各单元类型的堆叠。住宅设计规划（阳台、飘窗、花槽和悬架）和一层或二层高单元模块所用材料类型的变化的结合共同打造出住宅塔楼的立面图案。

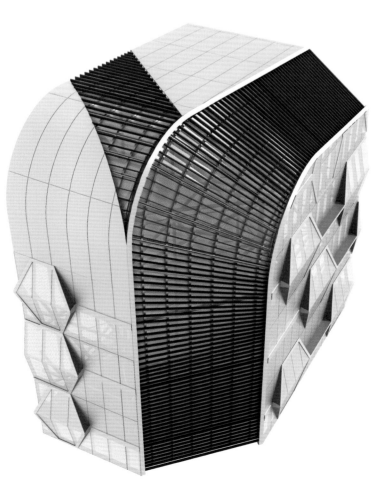

On the ground floor of the development stainless steel lines are inlaid into the floors and lines of light are traced across the ceiling, guiding pedestrians to their destinations. The design of the residential lobby uses the car park ramps to bring continuity from interior to exterior. The lines of light move from the exterior-covered walkway into the lobby, accentuating the ramps above, while the ceiling rises into a double height space at the elevators. The office lobby is divided into a reception area and a large café which extends along the view corridor to create a lively atmosphere in the public areas.

The sky lobbies and the sky gardens are an integral part of "V on Shenton" and provide 360-degree views of Singapore. The most ample and diverse of the three sky gardens covers the entire 8th storey of the development. Here residents are able to take full advantage of the amenities while still having privacy to train or entertain guests.

Along with the façades, the sky gardens are an integral part of developing the sustainable lifestyle of 'V on Shenton'. These lush green spaces provide a refuge from the city with the climate and vegetation, naturally providing fresher, cleaner air.

At the two sky lobbies in the heart of the residential tower, residents are given even greater privacy combined with views of the city or the ocean from both the 24th and the 34th storeys. The residents of the penthouse levels will also have exclusive access to the outdoor roof terraces on the 53rd and 54th storeys.

单元首层的地面上嵌着不锈钢线，天花板上有光标长排的灯光，指示行人前往目的地。住宅楼的大堂设计了车行坡道使内外连成一体。光标灯光从有蓬顶棚的人行走廊一直延伸进入大堂，突出了上方的斜坡，天花板在电梯口处向上升高了一倍。办公楼的大堂分成接待区和一个大咖啡区，咖啡区顺着观景走廊延伸，在公共区域内营造了一种活力气氛。

空中大堂和空中花园是珊顿五号项目里一处最主要的部分，这里可全方位地观看欣赏新加坡景观。三个空中花园里最丰富和多样的一个覆盖了住宅小区的整个第8层，在这里居民可充分享用各种便利设施，也可进行私人锻炼或招待客人。

塔楼立面连同空中花园是"珊顿大道五号"发展可持续生活方式的重要组成部分。这些丰富的绿色空间是城市里一处绿洲，让人们远离城市喧嚣和日晒雨淋，提供的自然植被则带来了新鲜清洁的空气。

在位于住宅塔楼中心的24层和34层的两个空中大堂，居民可更加自我地享受到城市或海洋景观。顶层的居民可独享第53层和第54层的屋顶平台。

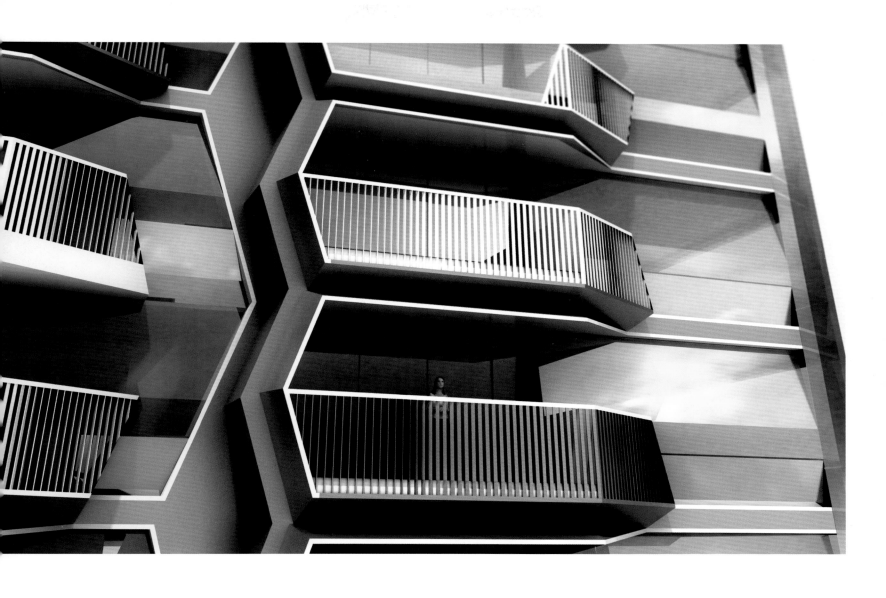

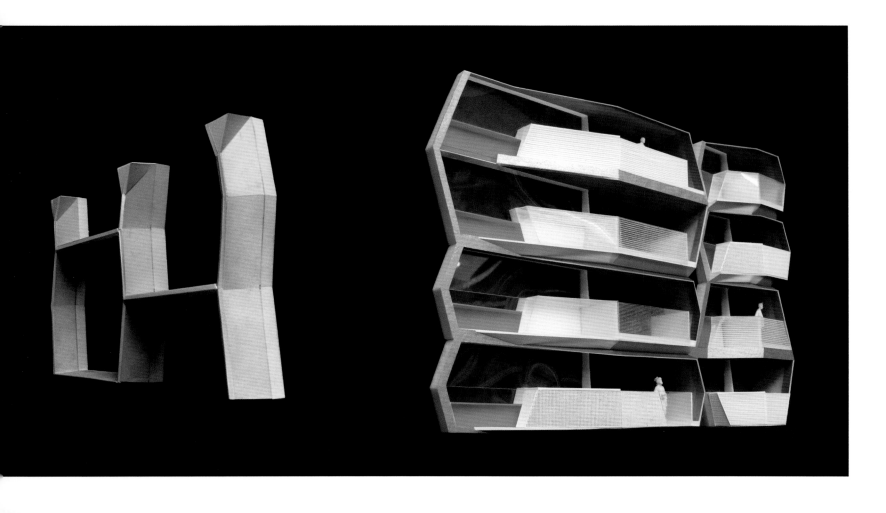

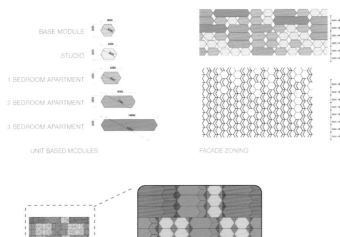

BASE MODULE

STUDIO

1 BEDROOM APARTMENT

2 BEDROOM APARTMENT

3 BEDROOM APARTMENT

UNIT BASED MODULES

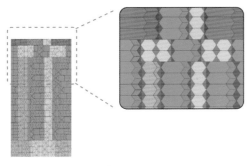

FACADE PATTERN SUPERPOSITIONED

FACADE ZONING

Scaler textures
肌理

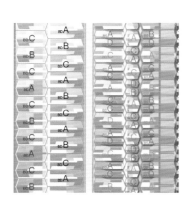

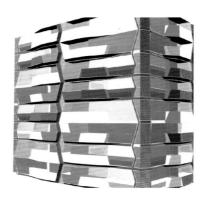

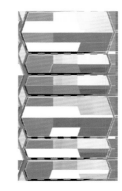

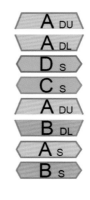

Residence Facade Pattern
住宅楼立面样式图

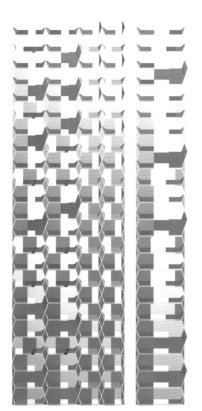

OPTION 01

OPTION 02

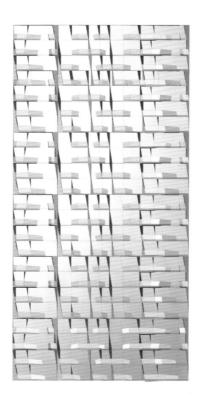

OPTION 03

Residence Envelope Pattern
住宅楼立面样式图

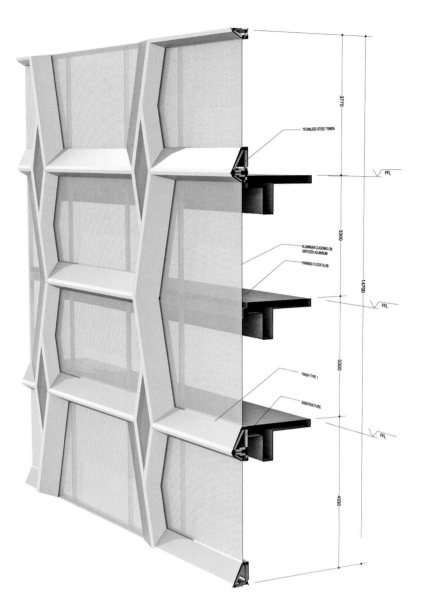

"STAINLESS STEEL" FINISH

ALUMINUM CLADDING ON UNITIZED ALUMINUM

PARKING FLOOR SLAB

FINISH TYPE 1

SUBSTRUCTURE

3770

3300

14700

3300

4330

FFL

FFL

FFL

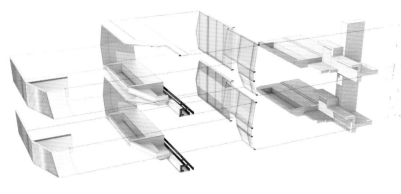

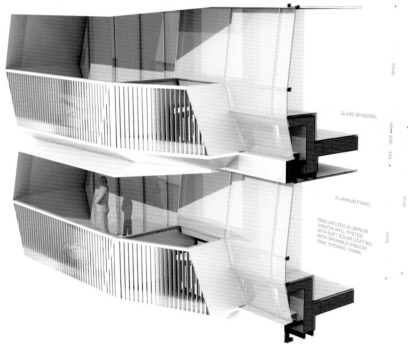

GLASS SPANDREL

ALUMINUM PANEL

SEMI-UNITIZED ALUMINUM
WINDOW WALL SYSTEM
WITH SOFT SOLAR COATING
WITH OPERABLE WINDOW
(MAX. OPENING 150MM)

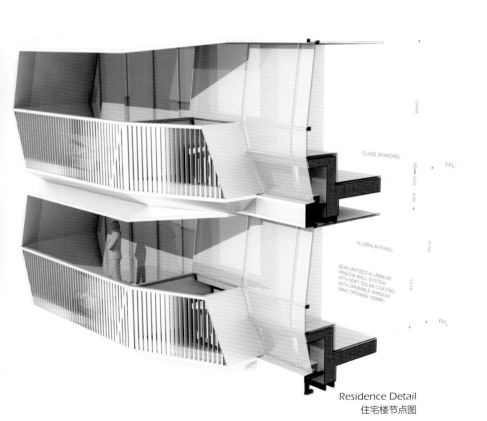

GLASS SPANDREL

FFL

ALUMINUM PANEL

SEMI-UNITIZED ALUMINUM
WINDOW WALL SYSTEM
WITH SOFT SOLAR COATING
WITH OPERABLE WINDOW
(MAX. OPENING 150MM)

FFL

Residence Detail
住宅楼节点图

Residence L24 Detail
住宅楼 24 楼节点图

MATERIAL 材料
FAÇADE 立面
SHAPE 造型
WINDOW 窗口

Sheung Wan Hotel,Hong Kong, China
中国香港上环宾馆

Architect: Heatherwick Studio
Location: Hong Kong, China
Photography: Heatherwick Studio

设计公司： 赫斯维克工作室
地点： 中国香港特别行政区
摄影： 赫斯维克工作室

Heatherwick Studio has been asked to design a 40-storey hotel with 300 rooms for a district of Hong Kong famous for selling unpackaged dried fish with stinky richly-textured seafood hanging from shop fronts and piled high in baskets. The atmosphere of streets like these can be wiped out when a row of small shops is replaced by a single flat, shiny building.

设计师被要求在香港著名的销售散装干鱼的地区设计一座有 300 个房间、40 层高的旅馆。这里的店铺挂满了充满臭气的海产品，篮子里也高高地堆满了海产品，当一排排小店铺被平整光洁的大楼所取代后，街道里的腥浊空气才能被清除干净。

特点分析

FAÇADE

The building's front façade is composed of thousands of "boxes".
These "boxes" protrude to different extents giving the hotel a very
different architectural texture from the smooth, shiny façades
normally seen on general buildings.

立面

建筑的外立面则由这些数千个不同表面的"盒子"所构成。这些伸缩凹凸
不等的"盒子"使旅馆建筑的结构非常与众不同，与那些拥有整齐平顺立
面的建筑完全两样。

Because most hotel projects deal with putting arbitrary new interiors into existing buildings, it is rare to find a connection between the inside and the outside of a hotel. This project, which was to build a hotel from scratch, is an opportunity to conceive the inside and the outside at the same time.

The idea that Heatherwick Studio developed was to interpret the familiar objects found in a hotel room—bed, window, mini-bar, safe and a place to keep the iron—as a series of boxes, of four different sizes. In every room, all the furniture and fittings are formed from a different arrangement of these boxes, making every room unique, and breaking down the scale of the new development so that it relates to the scale and grain of the existing street.

The building is a concrete structure, filled in with metal boxes, which are manufactured with the folded-metal technology used to make air conditioning ducts and water tanks. Inside the building, boxes are lined with bronze and sprayed directly with rigid insulation foam or upholstered to make beds and seats.

由于大多数旅馆都是在现有的楼房内随意改建而成，所以很难找到一个内外相通的旅馆。这个从零开始建造一座旅馆的项目酒店，是一次可同时兼顾到内外的绝好机会。

设计师的想法是重新诠释一个酒店房间里常见的物品：床、窗户、小冰柜、保险柜、放置熨斗的柜子等。每间房间的家具配置都依据这些铁盒子的不同安排而打造，使得每间房间都是独一无二的。新旅馆突破了新发展规划的范围，因此它关系到现有街道的比例和布局。

酒店建筑是混凝土结构，内嵌着用拆叠金属工艺制造的金属"盒子"，这种工艺用来制造空调管道和水箱。建筑内部这些"盒子"以青铜作内衬并直接喷涂上硬绝缘泡沫塑料或装上软垫用作床和椅子。

GREENERY 绿色
STRUCTURE 结构
SUSTAINABILITY 可持续性
BALCONY 阳台

The Scotts Tower, Singapore
新加坡史各士大厦

Architect: UNStudio
Client: Far East Organisation
Location: Singapore
Site Area: 6, 099.7 m²
Gross Floor Area: 18, 500 m²
Floors: 31
Height: 153 m

设计公司：UNStudio
客户：远东机构
地点：新加坡
占地面积：6 099.2 平方米
建筑面积：18 500 平方米
层数：31
高度：153 米

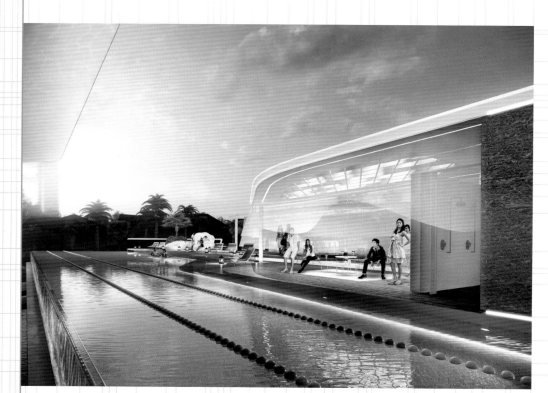

The Scotts Tower Soho apartment building in Singapore is situated on a prime location, close to the Orchard Road luxury shopping district and with views encompassing both nearby parkland and the panoramic cityscape of Singapore City. The design of the tower embraces both the neighbourhood principle and the history of the city of Singapore, alongside the hybrid conditions created by the prominent blend of architecture and nature inherent to the city.

史各士大厦坐落于新加坡的黄金地段，靠近乌节路豪华购物区，能在大楼上看见附近公园和新加坡全市市容。项目设计融合了邻里原则和新加坡历史，建筑与城市固有特征的出色结合而形成的混合环境相一致。

The four residential clusters are each designed for versatile and customised living. Individual identity is given to each unit by means of type, scale, distribution and articulation of outdoor space, along with the possibility for personalisation of the interior layout. The individual articulation of each cluster within the main framework of the tower is directly related to the organisation and materialisation of the terrace spaces. These varied outdoor spaces afford a choice of views, with corner terraces providing both cityscape panoramas and vistas over the natural landscape adjacent to the building.

The concept of the Scotts Tower is that of a vertical city incorporating a variety of residence types and scales. In addition, outdoor green areas in the form of sky terraces, penthouse roof gardens and individual terraces form an important element of the design. The vertical city concept is interpreted on the tower in three scales: the "city", the "neighbourhood" and the "home". The three elements of the vertical city concept along with the green areas are bound together by two gestures: the "vertical frame" and the "sky frames".

The project includes spacious landscaped gardens, sky terraces, roof gardens and numerous recreational facilities. The designer emphasizes that the most important thing is the interaction between the building and urban environment of Singapore. The project is not to design a simply horizontal layout, but to create a neighborhood up in the sky, a vertical city, where each space has its own unique identity.

四个住宅类型采用了通用的定制设计。用户可以自由选择每个单位的类型、大小、排列和链接，并进行个性化的室内装饰。同时，居民还可以根据每个楼台空间的组合和材料使用，遵从自己的意愿装点配置自己的生活。不同的室外空间提供了不同视野选择，每一个单元都有露台空间可以享受外面的自然风光，全市全景和建筑周边的自然景观。

史各士大厦的设计概念是用垂直形式来整合各种居住模式。此外，退台式户外屋顶花园、阁楼屋顶花园和个人活动绿地也是设计的重要元素。垂直城市的概念主要通过三个尺度得以诠释："城市"、"邻里"、"家"。这三个概念同绿化捆绑在一起，并通过"垂直框架"和"天空框架"完美地展现出来。

项目建设包含开阔的景观花园、天空露台、屋顶花园和众多的娱乐设施。主设计师强调这个塔楼重要的一点就是它要与新加坡城市环境互动。项目不是设计简单通常的水平布局，而是创造了一个天空中的邻里社区，一个垂直的城市，在这里每个空间都有自己独特的身份认证。

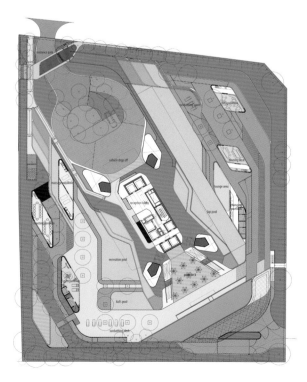

Landscape Plan
景观平面图

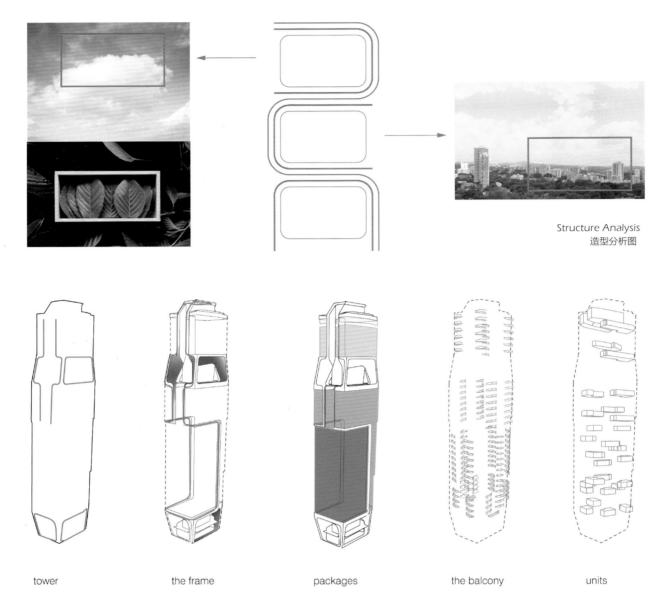

Structure Analysis
造型分析图

tower

the frame

packages

the balcony

units

Structure Analysis
结构分析图

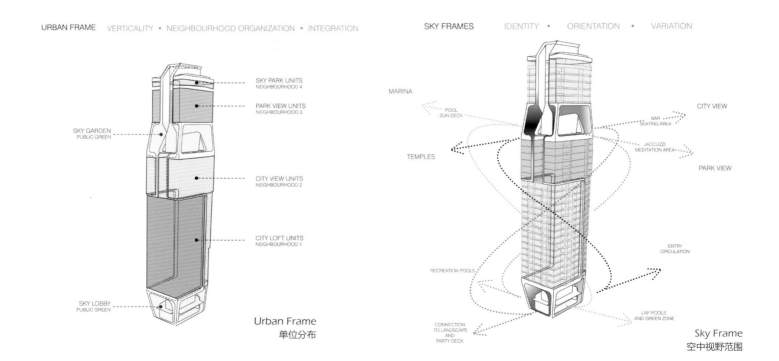

URBAN FRAME   VERTICALITY • NEIGHBOURHOOD ORGANIZATION • INTEGRATION

SKY FRAMES   IDENTITY   •   ORIENTATION   •   VARIATION

SKY PARK UNITS
NEIGHBOURHOOD 4

PARK VIEW UNITS
NEIGHBOURHOOD 3

SKY GARDEN
PUBLIC GREEN

CITY VIEW UNITS
NEIGHBOURHOOD 2

CITY LOFT UNITS
NEIGHBOURHOOD 1

SKY LOBBY
PUBLIC GREEN

Urban Frame
单位分布

MARINA

POOL
SUN DECK

TEMPLES

RECREATION POOLS

CONNECTION
TO LANDSCAPE
AND
PARTY DECK

CITY VIEW

BAR
SEATING AREA

JACCUZZI
MEDITATION AREA

PARK VIEW

ENTRY
CIRCULATION

LAP POOLS
AND GREEN ZONE

Sky Frame
空中视野范围

FEATURE 特点分析

STRUCTURE

The "vertical frame" organises the tower architecturally in an urban manner. The frame gives the tower—on a macro scale—the "vertical city" feel, while dividing the four residential clusters (packages) into different "neighbourhoods", which are identified through alterations in the tint of the glass. The micro scale in the design is provided by the balcony variations of the individual residential units, which provides the feeling of "home" (unit identity) to the residents. The "sky frames"— at the lobby (Level 1 & Level 2) and sky terrace (Level 25)— organise the amenity spaces and green areas of the tower. Situated above the lower sky frame, 128 City Loft residences in the first cluster occupy the lower 16 floors of the tower. Single urban unit with a multifunctional design, compact spaces and cutting-edge features, the City Loft residences measure 62 m².

结构

"垂直框架"组织展现了建筑的现代模式。从宏观角度看，这个模式赋予建筑一种"垂直城市"的感觉，汇集了 4 个不同的街区，并通过色调变换的玻璃来确定。从微观角度看，设计通过阳台的划分，赋予每个单元住户一种"家"的感觉。而"天空框架"空中花园在大堂 1—2 层以及天空露台 25 层，作为楼内的休憩空间和绿化地带。在低层天空框架空中花园的上方是 16 层拥有 128 个城市阁楼的住宅单位。单个占地 62 平方米的单元拥有多功能设计、紧凑的空间和前卫的设计特色。

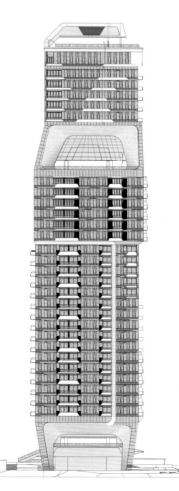

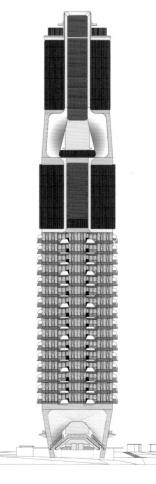

North Elevation
北立面图

East Elevation
东立面图

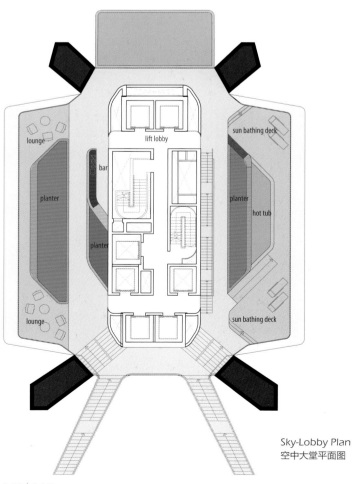

Sky-Lobby Plan
空中大堂平面图

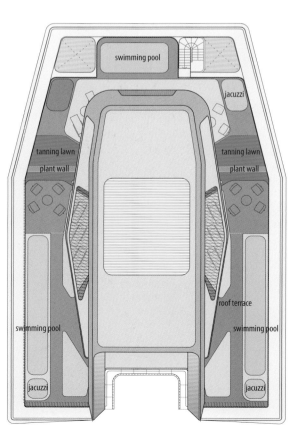

Roof Plan
屋顶平面图

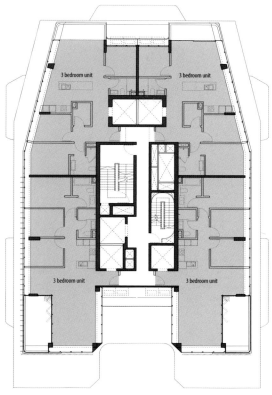

Floor Plan 3
楼层平面图 3

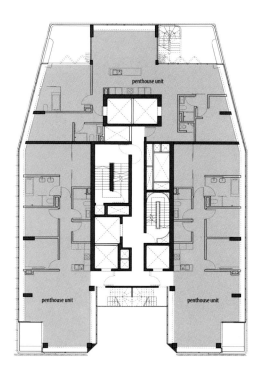

Floor Plan 4
楼层平面图 4

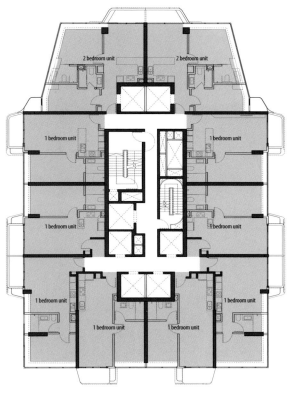

Floor Plan 1
楼层平面图 1

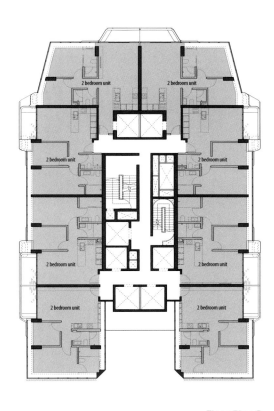

Floor Plan 2
楼层平面图 2

Park view units cluster above the second sky frame, covering five floors and containing twenty units. The park view urban family residences offer communal and retreat spaces in a motion friendly design, with each covering an area of 122 m².

The various landscape spaces consist of elements including planting, paving, and water features - including recreational waters, green waters, vapor waters, water walls and swimming waters. The landscape is articulated by two areas: urban recreation to the north and enclosure and relaxation to the south. Recreational facilities within the gardens include a 50 m lap pool with sunning deck, a children's pool, a wellness pool, dining & BBQ pavilions, a meeting pavilion, massage and gym pavilions.

拥有公园景观的公寓单元紧邻隔壁，是凌驾于第二组空中花园天空框架之上的下一组单元，有 5 层楼高，包含 20 个单元。这组景观城市家庭住宅每个单元占地122平方米，并通过运动友好型的设计提供了公共休闲空间。

多层次的景观空间包括植被、铺路和水景——休闲水域、绿化水域、蒸汽水域、水幕墙和游泳池。景观由两个区域组成，即北部的城市休闲区和南部的封闭消遣区域。花园里包含了各种娱乐设施，包括配有遮阳甲板的 50 米泳池、儿童游泳池、健身中心、餐厅、烧烤凉亭、会议区、按摩和健身区域等。

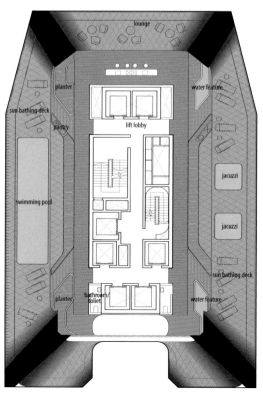

Sky-Terrace Plan
空中露台平面图

SHAPE 造型

FAÇADE 立面

VIEW 视野

BALCONY 阳台

Infinity, San Francisco, USA
美国旧金山无限大厦

Architect: Arquitectonica
Client: Tishman Speyer
Location: San Francisco, USA
Site Area: 92,903 m²
Floors: 37 (Main Street tower), 42 (Spear Street tower)
Photography: Joern Blohm

设计公司：Arquitectonica
客户：Tishman Speyer
地点：美国旧金山市
占地面积：92 903 平方米
层数：37（主街楼）、42（Spear 街楼）
摄影：Joern Blohm

The mixed-use residential project has four buildings–two low-rises and two high-rise–joined by public courtyards and sidewalks cutting through the centre and connecting it to the urban grid. The façades of the towers are soft and continuous curves of glass, giving multidirectional views with no front or back.

本案的多功能住宅项目共有四栋建筑——两栋低层和两栋高层，并且与公共庭院连成一体，人行道穿过中心并同城市网络相连。建筑柔和连续的玻璃曲线立面拥有多方向视野，通透无阻。

The program for this 92,900 m² mixed-use project consists of 91,057 m² of residential, 710 m² of amenities including a fitness centre and pool, 1,132 m² of retail, 767 parking spaces, public courtyards and sidewalks cutting through the block. They reach out to the corners, matching the streets and creating a central space of courtyards and walkways.

The design creats buildings that are transparent and light. Wherever you are, whether you're in the living room or in the bedroom, you sense that skin that is your exterior identity, reappearing inside. And whatever is done in the living room, particularly where the curve is more accentuated, it creates a panoramic view inside that living room. The living room, dining room and kitchen are conceived as a single space almost like in a loft, in order to take advantage of these views.

Infinity is the new landmark of San Francisco. The towers are transparent with interiors taking in the bay and the skyline. The towers flow around their corners with curves that relate to the waterfront location. No longer boxy, solid or massive. These are slender towers and each corner living room has the panoramic sweep of the view along its curving façade, like a monumental bay window. The kitchens are furniture pieces that open to the loft like open space.

这个占地 92 900 平方米的综合开发项目包括 91 057 平方米的住宅区、共占地 710 平方米的健身中心和游泳池、1 132 平方米的零售区、767 个泊车位、公共庭院和穿过其中的人行道。它们相互联系，与街道设置共同创建出融合庭院和人行道在内的中央空间。

建筑设计以透明光亮为主。无论你置身何处，客厅或是卧室，都能感觉到与自然的融合。无论你在客厅或是曲面感更强的地方做什么，都能坐享无限美景。客厅、餐厅和厨房被设想成一个单一的空间，就像都在一间阁楼里，如此一来就充分欣赏窗外美景。

无限大厦是一座旧金山新地标。透明的塔身、通透的视野、曲线的设计和海滨的位置，这些特点使得建筑不再四方、僵固和厚重。这些纤细的塔楼和坐落在每个角落的客厅都将沿着曲面的立面拥揽室外美景，如一个巨大的凸窗。厨房设置朝阁楼开放，看起来就像一个开放的空间。

FEATURE 特点分析

SHAPE

The design of the towers is based on a 4-sided cloverleaf. Each is wrapped in smooth undulating curvilinear glass surfaces, while the low-rise buildings below are rectilinear in plan to respect the urban grid. The expanses of glass would result in amazing, expansive views. It's about the boundary of inside and outside disappearing where owners will feel that they are out there in the middle of the sky with no framed views.

造型

建筑的造型设计灵感来源于四叶草的形状。每个建筑都有波浪起伏的光滑玻璃曲面，而其下部低层建筑的直线设计对应了城市网络。落地窗的设计更是亮点所在，极大地拓展了视野，柔化了建筑内外部的边界，让用户享受到凌驾于空中的感觉。

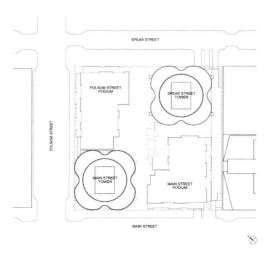

Site Plan
总平面图

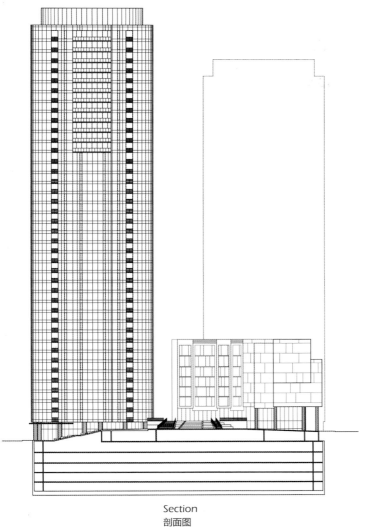

Section
剖面图

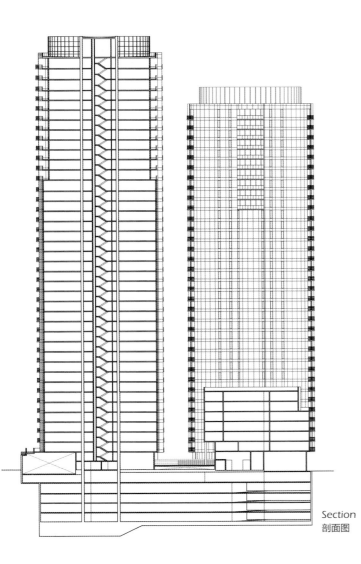

Section
剖面图

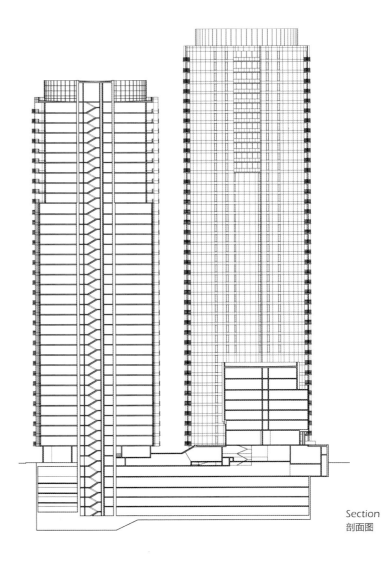

Section
剖面图

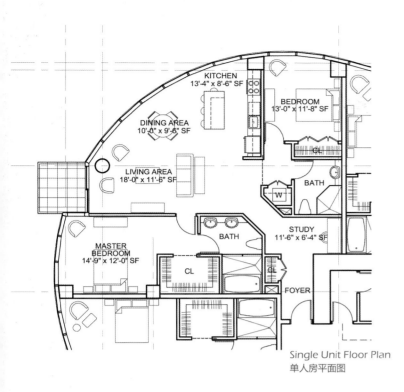

Single Unit Floor Plan
单人房平面图

Amenities are a big part of high-rise living. Think of a condominium like a small village with its community centre and village green. Infinity has a central courtyard that is the focus of the community. It has an indoor lap pool, amazing spa and fitness centre, party room, conference area and children's playing area. There are even shops and restaurants facing the street. So today a high-rise condominium is not just apartments on top of each other. It has amenities that one could not have in a single family home. Sharing has advantages. Plus, there is a lot to do in San Francisco.

便利设施在高层住宅中占据了很大一部分，使得公寓就像一个融合社区中心和广场在内的小村庄。项目设置一个中心庭院作为社区，有室内游泳池、绝佳的水疗中心和健身中心、联谊室、会议室和孩子的游乐区等，还有商店和餐馆临街而建。所以今天的高层公寓不仅仅是位于彼此之上的公寓。它拥有的设施是单一家庭所没有的，共享有优势。另外，在旧金山有许多可做的事。

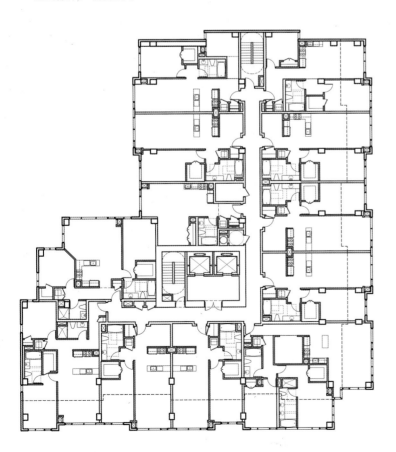

Typical Main Podium Plan
裙楼典型平面图

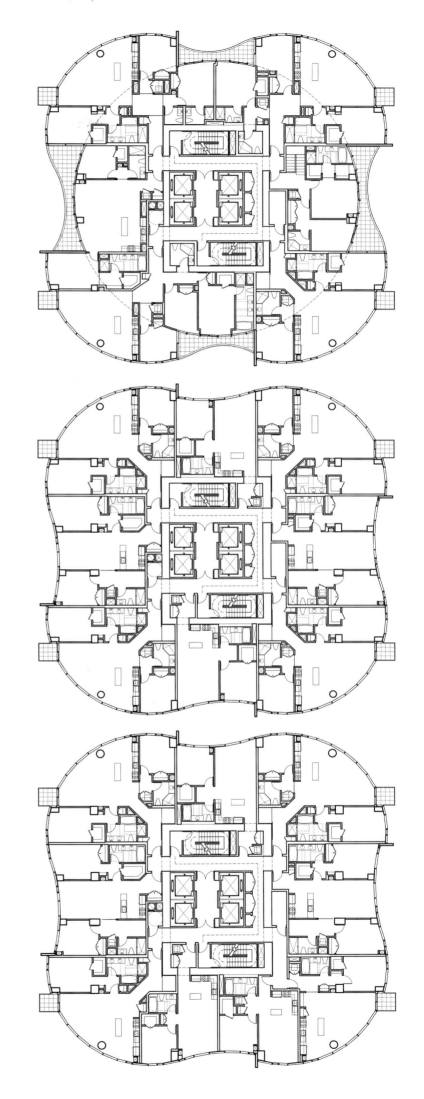

Typical Lower Level Plan
典型低层平面图

SHAPE 造型
BALCONY 阳台
FAÇADE 立面
MATERIAL 材料

The Helix
螺旋大厦

Architect: Zeidler Partnership Architects
Location: Abu Dhabi, UAE
Site Area: 80,000 m²
Statue: Under Construction
Renderings: David Liang

设计公司：Zeidler 及合伙人建筑事务所
地点：阿联酋阿布扎比市
占地面积：80 000 平方米
状态： 建设中
效果图：大卫·梁

The Helix is a twin residential tower development located on Al Reem Island, Abu Dhabi, near the waterfront beach, with clear views over the Central Park and the business district. These 31-storey condominiums adjoin a large 4-storey car parking element with street level retail and duplex residential units.

螺旋大厦是一座双子住宅大厦，坐落于阿布扎比的艾尔姆里岛，是滨水而建 31 层的高级公寓，可环视中央公园和商业区。大厦的首层集商场和复式住宅于一体，并邻接一座 4 层高的大型停车场。

BALCONY

Balconies that spiral out from the fully glazed facade are also rotated in plan, creating an upward spiral patterning on the façade. The almost continuous slab edge projection that forms the balconies also serves to shade the façade, reducing glare to the urban environment and solar gain to the interior spaces.

阳台

阳台以旋转式环绕着大楼的全玻璃立面，打造出一个螺旋式的上升模式。同时，构成阳台的连续突出厚板为立面起到遮阳作用，削弱了反射到四周的光线和直射到室内的阳光。

The roof of the parking podium is an extensive garden incorporating private terraces for residential units as well as public lounging areas for all the residents around a health club pavilion and swimming pool areas. The swimming pool is visually extended as a shallow reflecting pool along much of the façade that faces the beach.

The façades of the parking structure along the collector road are treated as sculptural architectural elements mixing natural stone, with metal accent screens. The screen elements are used to hide traditional parking garage language, enhancing the image of the project for both pedestrians and the surrounding neighborhood.

停车场的屋顶被设计成为一座宽广的花园，不仅能作为各住宅单元的私人露台和公共休息区，而且设置有健康俱乐部和游泳池以供人们休闲娱乐之用。游泳池在视觉上向外延伸，酷似一个浅倒影池，形象地映射出面对着海岸一侧建筑的风采。

停车场沿着次干路的立面构造充满了雕刻般的建筑元素，融合了天然石材和金属网屏。网屏的元素取代了传统室内停车场的建筑语言，从而增强了行人和周围街区对建筑的整体感认知。

The architectural design of the complex reflects the desire to make the most of this unique site. Elliptical towers are placed at opposing angles to create a dynamic massing from all directions. The floor plates open panoramic views from all condominium units. The units themselves are a mixture of studios and 1 to 2 bedrooms, all with contemporary kitchen and bathroom designs.

建筑的设计反映了设计师的愿望，即建造出该地区最独一无二的建筑。两座椭圆形双塔相向而立，从各个方向打造出动感的体块。开放式楼层布局让所有房间都能欣赏到开阔的全景。每一房间都拥有工作室和一到两间卧室，并配置了现代化的厨房和浴室设计。

SUSTAINABILITY 可持续性

SHAPE 造型

FAÇADE 立面

WINDOW 窗

Collection House for Scientists in Guangzhou Science City, Guangzhou, China

中国广州科学城科学家集合住宅

Architect: AXS Satow Inc., The Architectural Design & Research Institute of Guangdong Province
Location: Guangzhou, China
Site Area: 39,957 m²
Gross Floor Area: 105,407m²

设计公司：株式会社佐藤综合计画、广东省建筑设计研究院
地点：中国广州市
占地面积：39 957 平方米
建筑面积：105 407 平方米

This is a residential building for the scientists working on projects overseas in order to develop the Science Park. It has the function to use the prevailing wind as regional characteristic to prevent the subtropical heat and rain.

为促进科技园发展，满足海外科学家的住宿要求，广州启动了科学城科技人员公寓的建设。建筑充分利用当地盛行风特征来防止亚热带热辐射和降雨。

FEATURE 特点分析

SHAPE

The shape of the building is designed according to the local climatic conditions. Innovative oval shape and the traditional rectangular shape are organically combined, responding to local prevailing wind and sunshine conditions. The bottom of the building has big space. Multilayer plate apartment breaks the closed interior space, forming a natural ventilation pattern.

造型

建筑的造型是根据当地的气候条件而进行设计的。新颖的椭圆造型和传统的长方形造型有机地进行结合，顺应当地盛行风向和日照进行布局。底层架空，板式多层公寓打破空间的封闭，形成自然的通风格局。

The vertical louver in each unit controls light from sun and also works as a screen to block people's eyes but still has high transparency. The sky court appearing in the centre of elevation of residential tower is located at every six floors and functions as a green window in order to let wind in from outside.

The layered green-terraces are open spaces between each unit, which help residents to communicate with neighbors. Because the green comes into the piloti in the first floor, it gives extra visual sense of unity with the surrounding green space and useful essence as the activity along the building.

每个单元配置有垂直天窗，不仅可以调节阳光，而且可以作为一种屏障遮挡外部的视线，但同时也具有较高的透明度。住宅楼每六层中央位置会设有一个天窗，作为一个绿色窗口让风从外部吹入室内。

每个单元之间的绿色露台是一个开放的空间，方便邻里之间的沟通。由于一层架空柱引入了绿色元素，使得它作为一个延续部分为周围绿地和体量提供统一的视觉感。

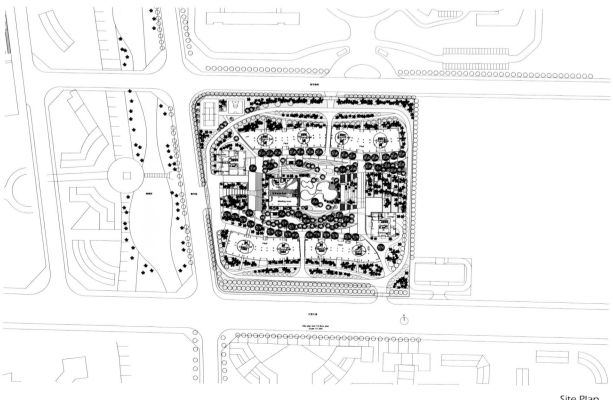

Site Plan
总平面图

At the connection between the four lower units, there is void space which obtains the "wind path" in the first floor. This "wind path" works as the extension of living space and also controls severe heat and rain as a semi-outdoor space.

Community garden is a leisure space full of green. Lotus pond, in the middle, is a popular place for people to relax and stroll. At the same time, fully functional facilities meet residents, daily needs, in order to provide a good environment and a comfortable living space.

在底层 4 个单元之间的连接处设有一个孔隙空间，允许风从一层通过。这种通风路径可以看作是居住空间的延伸，并作为一个半室外空间调控热量和雨量。

社区花园是一个充满绿色的休闲空间。荷花水池置于中间，是人们放松、漫步的好去处。同时，齐全的功能设施满足居民生活需要，为人们提供了良好的环境和舒适的居住空间。

Lowrise Longitudinal Section
低层纵剖面图

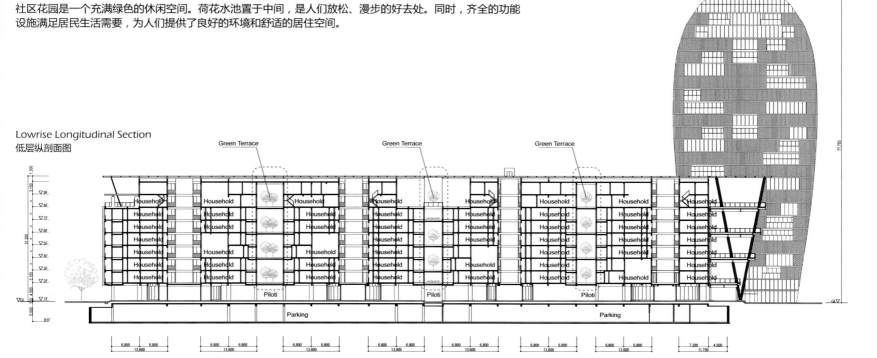

BALCONY 阳台

FAÇADE 立面

STRUCTURE 结构

SHAPE 造型

Changjing Zhige ABC1 Lands Mixed-use Development in Wuhu, China

中国安徽芜湖长江之歌 ABC1 地块城市综合体

Architect: Jiang Architects & Engineers
Client: Wuhu Guodu Real Estate Co., Ltd.
Location: Wuhu, Anhui, China
Gross Floor Area: 2,260,000 m²

设计公司：上海江欢成建筑设计有限公司
客户：芜湖市国都置业有限公司
地点： 中国安徽芜湖市
建筑面积：2 260 000 平方米

The project Changjiang Zhige is located in Wuhu City, Anhui Province, which is nearby shore of Yangtze River, with total land use of 580,000 m² and gross floor area of 2,060,000 m². It includes several functions, such as residence, hotel, business, office, etc. In the first phase, land A is developed, where use area is 152,000 m² and constructing area is 590,000 m². The main function of the first phase's buildings is high-rise residence.

长江之歌项目位于安徽省芜湖市，滨临长江水岸，总用地 58 万平方米，总建筑面积 206 万平方米，主要功能有住宅、酒店、商业、办公等。一期实施 A 地块，用地面积 15.2 万平方米，建筑面积 59 万平方米。一期主要功能为超高层住宅。

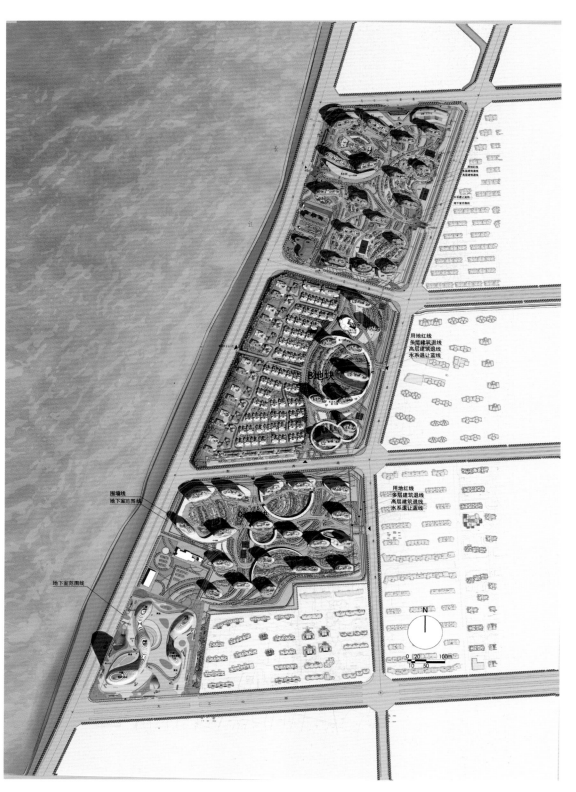

The basement is full basement, with a constructing area of 190,000 m² and a use area of 100,000 m². The east–west length is around 290 m and south–north 470 m. The design of the ceiling's draining system is difficult. To avoid punching in ceiling, we take plastic drainage board as a draining blanket. Surface water will flow through drainage board and into drainage ditch out of the ceiling board in order.

To save energy, the basement takes a light-guide system, which will introduce natural light to inner system. By strengthening effective transmission through light-guide system, diffuse reflector introduceds adequate light into the inner space.

90% shear wall is vertical of sole plate without any change; 10% shear wall in both sides gently forms an inclined wall with architectural pattern. Hence, the horizontal pull caused by inclined wall will be undertaken by floor, so that reinforcing bar area increase.

The concrete frame of 100-m roof takes a structure of bi-directional plane arch + empty horizontal beam structure, to decrease the influence caused by wind load on the whole structure.

地下室是满堂地下室，建筑面积 19 万平方米，用地面积 10 万平米。东西向长约 290 米，南北向约 470 米。地下室的顶板排水成了设计中的一个难点，为了不在地下室顶板上开洞，我们采用了塑料排水板做排水层，通过排水板的导流设计，将地面水有组织地流到地下室顶板外的排水沟内。

为了节约能源，地下室采用了光导照明系统，通过这种装置将室外的自然光导入系统内部。经过光导装置强化并高效传输后，由漫反射器将自然光均匀导入室内。

约 90% 的剪力墙垂直于基础底板，没有转换；在平面两端的 10% 的剪力墙为配合建筑外形和顺包络成为斜墙，斜墙变斜率处因此出现的水平拉力由楼板承担，板配筋相应增加。

百米屋顶混凝土构架采用双向平面拱 + 空腹水平梁结构体系，最大程度减小风荷载对整体结构的影响。

Site Plan
总平面图

In this project, extra large-scale garage is designed, and indoor ventilation becomes a design difficulty as well. In this case, intelligent ventilation system will be taken, namely ventilation will be divided into two processes of diluting and discharging.

In the diluting process, one feature of intelligent ventilation system is even distribution in garage. Besides, added receptor of pollutants and other controlling equipment are also designed. When the rate of pollutants exceeds standard in local place, ventilation equipment in local area can dilute concentrated pollutants in a quick speed by taking advantage of its feature of high-speed jet-flow.

In the process of discharge, there will be a period of dilution. Then, when the receptor of intelligent ventilation system and other equipments show a requirement of full ventilation in terms of control strategy, it means that the rate of pollutants is high so that space and dilution can't go ahead as normally. Now, it is time to start ventilation equipments and discharge pollutants.

This strategy of "diluting firstly and then discharging" can greatly decrease the use of equipment and ensure air quality as well, which both saves energy and increases equipment's service life.

本项目设有超大型地下汽车库，汽车库的室内通风问题也成为设计难点。设计采用智能型诱导通风系统，把汽车库通风换气分成先稀释、后排放两个过程。

在先稀释的过程中，智能型诱导通风系统具有在车库空间内分布均匀的特点。利用设备上加装的污染物质感受器及其控制组件，当局部点的污染物质超标时，可由局部点的诱导通风设备利用其高速喷流的扰动特性，快速稀释局部点较为集中的污染物质。

在后排放的过程中，经过一段时间的稀释混合，当智能型诱导通风系统的感受器及其控制组件对照控制策略判断需要全面通风时，说明整个通风换气空间的污染物质浓度已经较高，已无法实现将污染物质继续吹散稀释的作用。此时再开启全部的通风设备，将污染物质排出室外。

此种"先稀释后排放"的控制策略，在保证整个汽车库空间的空气质量的前提下，大大减少设备的开启时段，节约了电能，增加了设备的使用寿命。

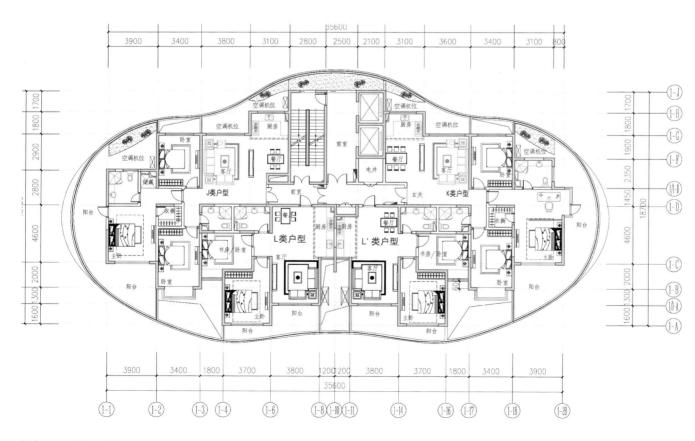

S1 Standard Floor Plan
S1 标准楼层平面图

FEATURE 特点分析

FAÇADE

In this design, the design of out façade is a main feature. It takes open balcony as horizontal element and emphasizes irregular plane curve, to create a straight and brilliant façade. This change brings irregular element to plane and profile, such as misaligned balcony, wall slanted inward or outward. Misaligned plates in balcony make it difficult to fulfill separation and burglarproof issue in balcony. The design takes brick masonry wall and bi-directional dull-polished glass, which protects residents, private space and decreases the rate of stretching across among residents.

立面

外立面是设计的主要特点，立面设计采用开敞阳台为横向线条的元素，强调不规则的平面曲线，使立面造型挺拔而炫亮。由于立面上的这种变化，平面和剖面上产生了不规则的元素，如上下不能重合的阳台板、向内或向外倾斜的墙体等。阳台上下两块板不能对齐使阳台分户墙的分隔和防盗的问题难以解决。设计采用了砌体墙加双向磨砂玻璃的做法，既保护了住户的私密空间，也降低了相邻住户间横跨翻跃的几率。

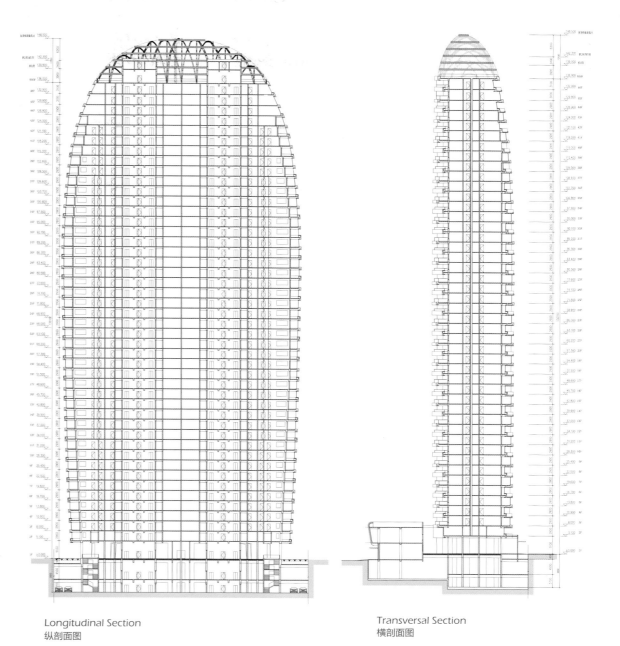

Longitudinal Section
纵剖面图

Transversal Section
横剖面图

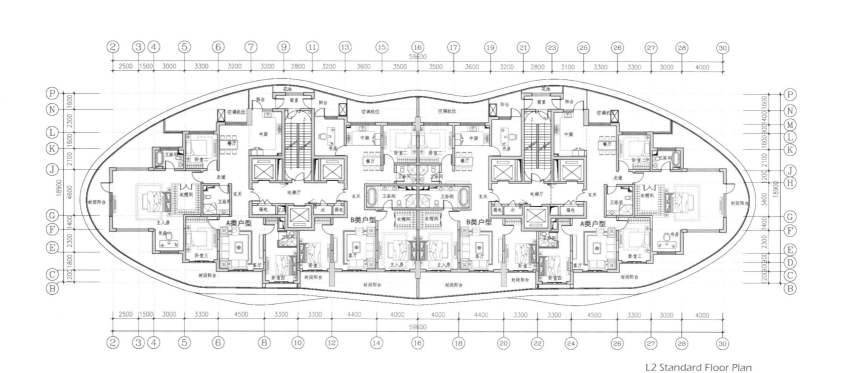

L2 Standard Floor Plan
L2 标准楼层平面图

The area is relatively large, with east-west length of 290 m and south-north length of 470 m. In southern and northern part of the second underground, small pump houses are equipped for water use in different places, to fulfill balance and decrease energy consumption.

The building's floor is designed between 33 and 44. Fully considering energy conservation and load of tubes, two water supply methods namely variable frequency and high cistern are taken in two areas divided by the height of 100 m. In southern area and northern area, living water is supplied by water pool and variable frequency pump package together between 6th floor and 3rd floor (below 100 m). The use of variable frequency pump package aims at saving energy, where compress will not be decreased. As to the part above 100 m, roof tank is taken to supply water.

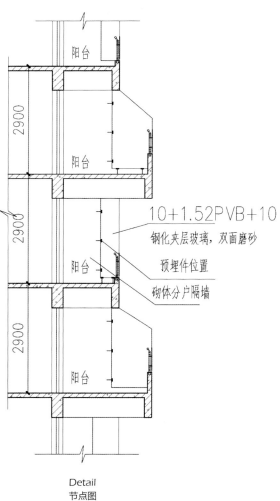

阳台

2900

阳台

2900

10+1.52PVB+10
钢化夹层玻璃，双面磨砂
预埋件位置
砌体分户隔墙

阳台

2900

阳台

Detail
节点图

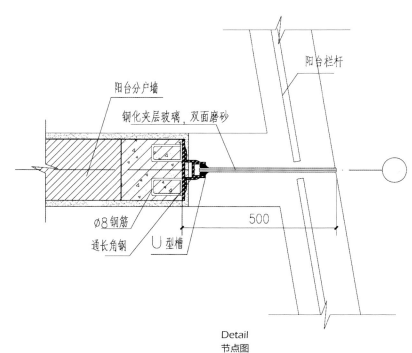

阳台分户墙

钢化夹层玻璃，双面磨砂

阳台栏杆

ø8钢筋
通长角钢
U型槽

500

Detail
节点图

因基地面积较大，东西向长约 290 米，南北向长约 470 米，地下二层南北面分设生活泵房供南北不同区域用水，可供水均匀并减少能耗。

此地块高层建筑为 33—44 层不等，从节能角度出发并考虑到管道不宜长期承压太高，以 100 米为界上下区域分别采用变频及高位水箱供水这两种不同的供水方式。南北区 6 层至 33 层（100 米以下）生活给水均由水池和变频泵组联合供水，使用变频泵组时为充分节能，每组变频泵的供水区域内不再减压。100 米以上部分采用屋顶水箱重力供水。

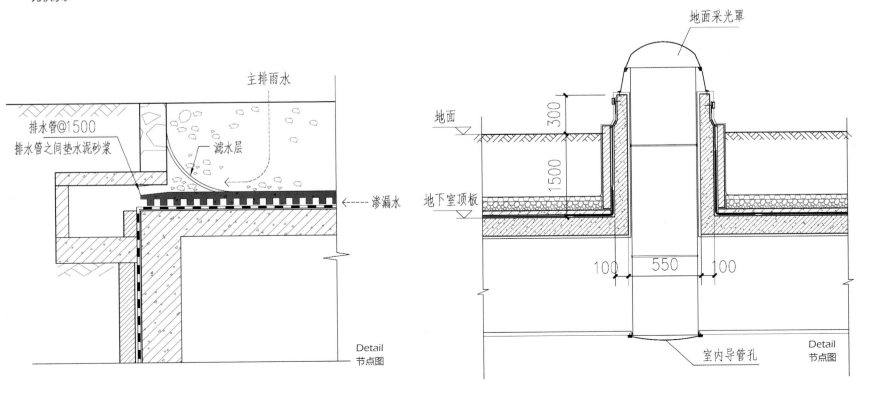

ENERGY SAVING 节能

SUSTAINABILITY 可持续性

SEISMIC 抗震

STRUCTURE 结构

Taipei Bade Urban Renewal Residence, Taiwan, China

台北八德路都市更新住宅项目

Architect: Arup
Client: Taipei Bade Urban Renewal Development
Location: Taipei, Taiwan, China
Gross Floor Area: 52,500 m²
Height: 90 m
Images: TLDC

设计公司：奥雅纳工程顾问
客户：台北八德路城市重建住宅项目发展中心
地点：中国台湾台北市
建筑面积：52 500 平方米
高度：90 米
图片：TLDC

As the city of Taipei enters a new era of urban renewal, the Bade Residences set a new benchmark for low-carbon living and pioneer the concept of "green-luxury-residential buildings" in Taiwan.

随着台北市进入城市更新的新时代，八德住宅区已成为低碳生活的新标杆，或为台湾"绿色豪华住宅"概念的先行者。

Arup is leading the designer for the Taipei Bade Urban Renewal Residence as part of an urban renewal scheme in Taipei. With natural ventilation, ample daylight and numerous sustainable amenities, this project aims to set the benchmark for sustainable design in Taiwan's residential market.

The project consists of four residential towers, a podium for commercial areas and amenities and a three-level basement with public parking and a swimming pool with a gross floor area of 52,500 m² above ground.

The project incorporates other sustainable building systems such as water-cooling centralised chiller tower, outdoor air (OA) system with heat recovery, rainwater and greywater recycling and automatic waste collection system. Building systems and façades are designed with exemplary energy performance targets in mind.

Each of the four towers features panoramic views of the city and landscaped terraces, which takes advantage of the centrally located site and the unusually high 90 m height restriction. The location of the towers has been carefully studied and guided through microclimate analysis so residents will enjoy optimum sightlines of the surroundings. Most of the towers are crowned by sky lounges and crystalline penthouses which provide luxury living spaces for residents to enjoy the view.

Unlike the traditional Taiwan approach to spaces, the floor plan is a decentralised living space extended outwards. This arrangement allows optimal unit size and proportions to encourage natural ventilation and light penetration. The living spaces are bathed in natural light with floor to ceiling windows providing panoramic views of the city.

台北八德路都市更新住宅项目是台北市重建计划的一部分，由奥雅纳工程顾问领衔设计。拥有自然的通风条件、充足的光线和大量可持续建筑设施，项目力求为台北的住宅市场订立一个可持续设计的标杆。

项目地上总建筑面积为 52 500 平方米，包括四栋 90 米高的住宅大楼、用作商业目的的区域和便利设施，以及用作公共停车场并设有一个游泳池的三层地下层。

建筑物整合了可持续运营系统，包括雨水和灰水回收设施和自动垃圾收集系统、带有热量回收的新风系统，以及一个集中型水冷式冰水机系统。建筑系统和立面设计也以节能为目标。

利用中央地段及不同于寻常的 90 米限高这两个优势，站在这些高层住宅楼顶风景优美的平台，台北全城景色尽收眼底。这些高层住宅楼的位置是在规划过程中采用微气候分析方法谨慎研究确定的，在保证最佳视野的前提下充分考虑了有效的通风与散热舒适性。大部分高层顶部有中央休息室和透明阁楼，为居民提供了观景和奢华生活空间。

与台湾传统的内向型高层建筑设计不同，这四座高层建筑的特色是全新的外向型设计，使居住空间更趋分散。这种独创性的布局可获得最理想的单元空间及比例分配，以促进自然通风和采光的使用。整个单元沐浴在自然光线中，通过落地窗可将城市美景尽收眼底。

FEATURE 特点分析

SEISMIC

Taiwan is a highly seismic zone. As such, the structural design of the project has incorporated a reinforced concrete shear wall and moment frame system with Steel Reinforced Columns (SRC), voided slab system and dampers.

抗震

台湾是一个地震高发区域。以此，项目的建筑结构符合抗震规范，采用了钢筋混凝土剪力墙以及有阻尼系统钢骨结构。

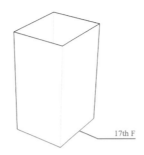
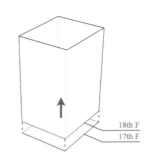
17th F
18th F
17th F

鑽石型頂部位
Diamond top

Structure
结构图

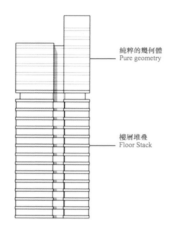
純粹的幾何體
Pure geometry

樓層堆疊
Floor Stack

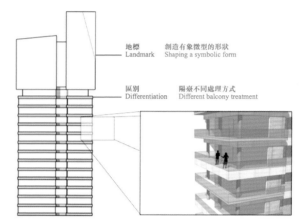
地標 Landmark　創造有象徵型的形狀 Shaping a symbolic form
區別 Differentiation　陽臺不同處理方式 Different balcony treatment

Facade
立面图

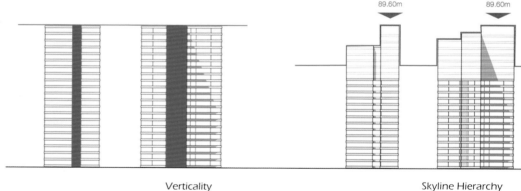
89.60m　　89.60m

Verticality
垂直感

Skyline Hierarchy
天际线层次分明

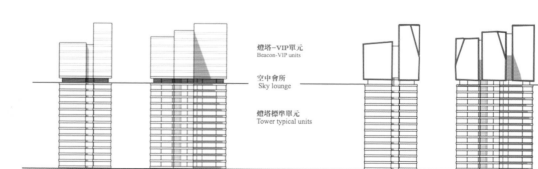
燈塔-VIP單元
Beacon-VIP units

空中會所
Sky lounge

燈塔標準單元
Tower typical units

Functions Distribution
功能分布图

ECOLOGY 生态

FAÇADE 立面

SUSTAINABILITY 可持续性

SHAPE 造型

Levent Loft II, Istanbul, Turkey

土耳其伊斯坦布尔乐文特阁楼花园 II

Architect: Tabanlioglu Architects
Client: Akfen GYO & Saglam Construction
Location: Istanbul, Turkey
Site Area: 1,759 m²
Gross Floor Area: 22,500 m²
Floors: 21
Height: 91.37 m
Photography: Helene Binet

设计公司：Tabanlioglu 建筑事务所
客户： Akfen GYO & Saglam Construction
地点： 土耳其伊斯坦布尔市
占地面积： 1 759 平方米
建筑面积： 22 500 平方米
层数：21
高度：91.37 米
摄影：Helene Binet

In the hearth of Istanbul, a 10,000 years old habitat, yet a city of youthful dynamism with its eye on the high-tech future and innovations, Levent Loft project faces requirements of the today's city-life within the concept of loft.

乐文特阁楼花园项目位于伊斯坦布尔的边界。伊斯坦布尔一处拥有 10 000 年历史的古老圣地，也是一个年轻又有活力的城市，正朝着高新技术发展之路奋勇前进。乐文特阁楼花园项目便在阁楼理念融入现代城市生活的需求下应运而生了。

FAÇADE

The mid-rise transparent residence block, with its vertical gardens, will emerge at the back end of Levent Loft I—a narrow horizontal structure to create another cohesive and lively alcove. Comprised of loft apartments at various sizes and forms, with large wall of windows and high ceilings, the residential building offers alternative layouts for distinctive requirements.

立面

建筑中高层的透明住宅区域伴随着垂直花园，将会出现在乐文特阁楼花园Ⅰ的后部——一个狭小的水平结构——形成另一个兼具凝聚力和活力的空间。建筑由不同规模和形式的阁楼式公寓组成，配合大型玻璃墙和高高的天花板，并提供了各样式的布局以适应不用的需求。

Following Levent Loft I, a transformation project on Maslak-Levent axis, at the CBD of Istanbul, Levent Loft II is designed in the concept of "soft loft".

In the understanding of loft style, exposed ductwork, plumbing, beams, concrete flooring, masonry and corrugated steel elements are kept in the spirit of openness and creativity that are integral to loft design.

Still, it is able to offer more conventional condo-like feel and to provide more energy efficiency, and such basics of loft are preferred with softer edges, including partial drywall encasements. Upscale loft apartments have defined spaces, especially bedrooms and fine flooring and they feature high-end kitchen and bathroom fixtures and finishes whilst the key elements of openness and versatility in loft design are respected.

The 21-storey building is nature-friendly with private walkout terraces, balconies and garden patios. The loft apartments have floor-to-ceiling, one-piece, energy-efficient windows open to the Bosphorous and the city panorama; wealth of natural light creates more commodious and ambient spaces.

继乐文特阁楼花园Ⅰ之后，乐文特阁楼花园Ⅱ作为一个转换项目，以"柔和的阁楼"为设计理念，坐落于伊斯坦布尔的CBD。

建筑设计诠释了阁楼风格，暴露的管道系统、管道、横梁、混凝土地板、砌筑和波纹钢元素充分延续了设计开放性和创造性的理念，这些都是阁楼设计不可或缺的元素。

尽管如此，建筑却能够提供常规的公寓感觉，并提高能源利用率，这样的基础配合柔和的边缘及局部干式墙包装成为了阁楼建造的首选条件。高档阁楼式公寓已经对空间做好了定义，尤其是卧室和上等地板，设计高端的厨房和卫生间装置。同时，开放性和多功能性也成为阁楼设计受人喜爱的关键性因素。

21层的建筑融合私人露台、阳台和花园楼台在内，是一个环境友好型住宅建筑。阁楼公寓拥有从天花板连通到地面的节能窗户，朝向博斯普鲁斯海峡打开，可以纵览城市全景。丰富的自然光线创造了更为宽敞舒适的空间。

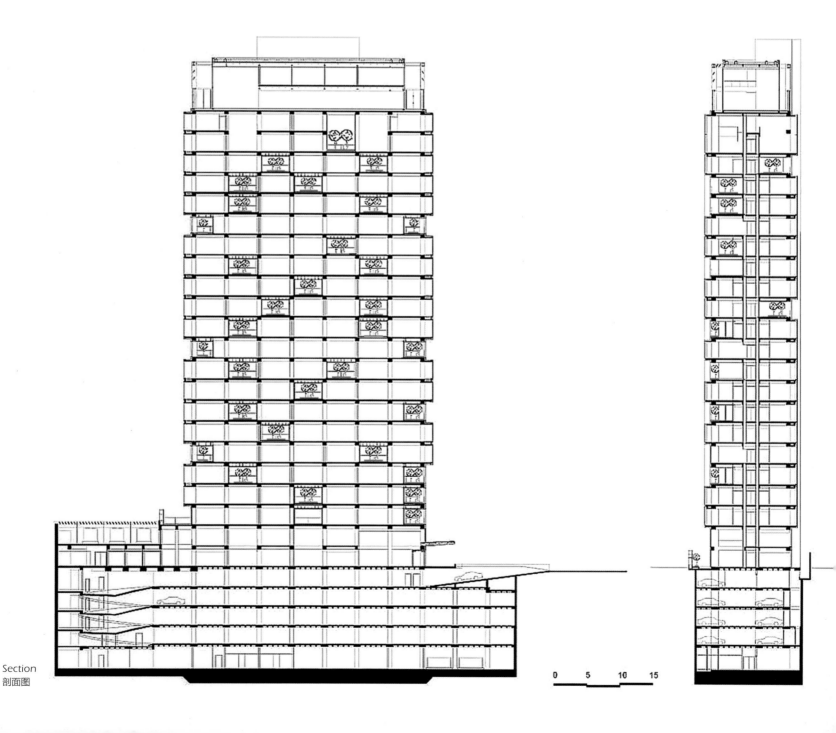

0   5   10   15

Section
剖面图

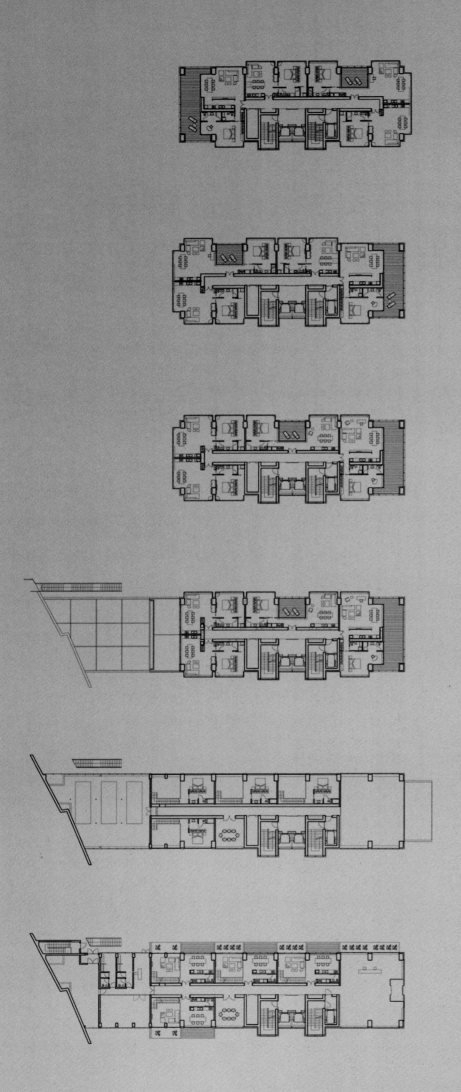
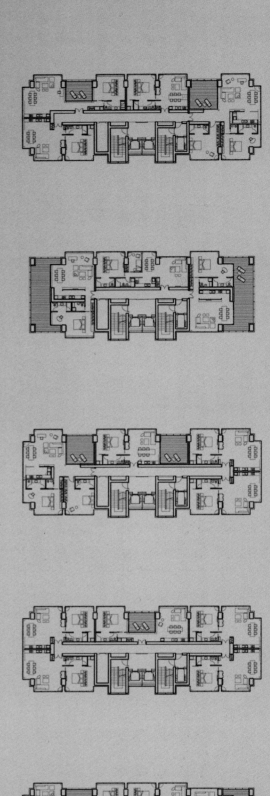
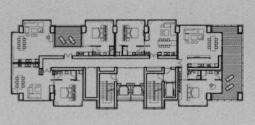
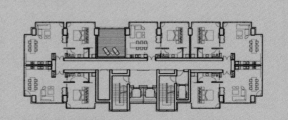

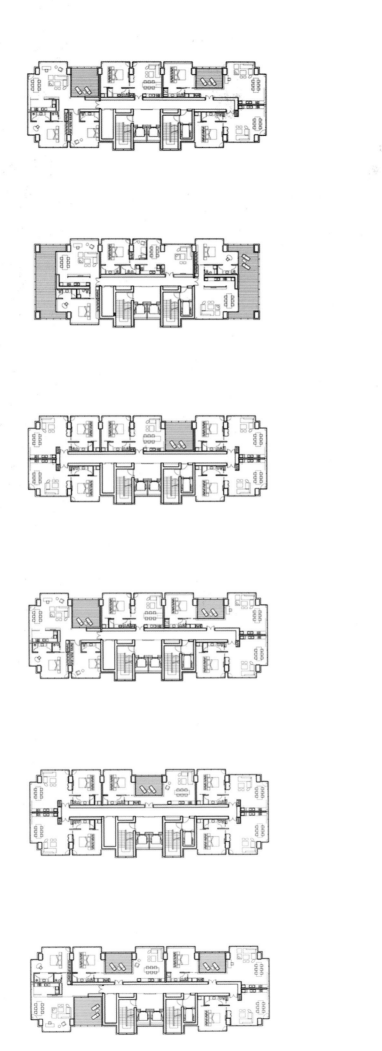
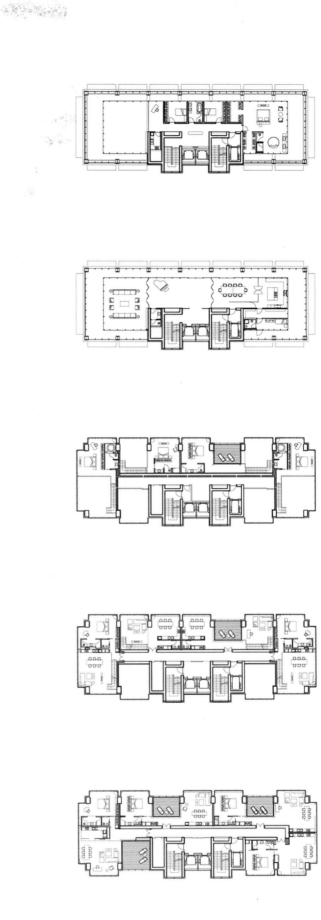
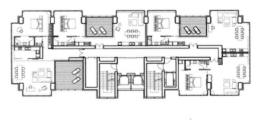
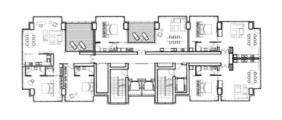

Floor Plan
楼层平面图

Facilities like the large underground car park, personal storage, central heating and ventilation, security, maintenance services and "intelligent house" system grant a user-friendly environment and easy access. The complex, supported by the recent technologies and professional service administration, offers a comfortable and fashionable life-style.

Levent Loft II is a step to shopping, cafes and restaurants; the subway stop is only one block away; the loft's life style is convenient especially for hardworking young people in search of a more liberal living form, who need to save time at home for social activities.

大型地下停车场、个人仓库、中央暖气和通风系统、安全服务、维护服务和"智能的房子"系统等设施创造了一个方便快捷的环境。这处综合建筑以最新的技术和专业的服务管理为支撑，为人们提供了一种舒适又时尚的生活方式。

乐文特阁楼花园 II 是用来解决购物、咖啡和餐饮的一种措施，与地铁车站只隔一个街区。阁楼生活方便快捷，为辛勤工作的年轻人提供一种更加自由的生活形式，以便节省在家时间用来参与社交活动。

MATERIAL 材料

FAÇADE 立面

SHAPE 造型

Bay-window 凸窗

1 Unit • 100 Families • 10,000 Residents, Shenzhen—Happy Town

一户·百姓·万人家——深圳幸福城

Architect: WSP Architects
Client: Urban Planning Land and Resources
Commission of Shenzhen Municipality
Location: Shenzhen, Guangdong, China
Site Area: 41,694 m²

设计公司：维思平建筑设计
客户：深圳市规划和国土资源委员会
地点：中国广东省深圳市
占地面积：41 694 平方米

The project takes efforts to integrate landscape into the city maximally to form a close relationship and an intimate planning. It primarily focuses on traffic-oriented and living environment improvement. Also named the "Happy Town", the project as a representative of an ideal city, consists of 2 levels of underground parking, traffic space, ground floor, an open commercial block in 2 levels, a private rooftop garden, empty ground spaces and upper residence, which together will form a mixed urban living circle. An open central square is indispensable in a community, so does the project. The square will be a multi-functional one for various events and activities. It can be a market in the morning, food stalls in the evening and celebrations in festive days.

万人家的规划上致力于将景观最大限度地融入城市，形成独特的图底关系和规划特征，主要关注于城市的疏导和生存环境的改善上。幸福城作为理想城市的代表，由 2 层地下停车及交通空间、首层与 2 层开放式街区商业空间、私密的空中花园、架空的底层以及上部的高层住宅组成，形成一个混合都市生活圈。在这样的保障房社区内，开放的中心广场不可或缺。它的功能可以是多元化的，目的只是为来自城市和生活在这里的人们提供聚集和发生各种活动的场所。清晨这里可以是早市、菜市场，晚上商店关门后这里又是大排档的美食广场，节庆的日子这里也是市民欢庆的场所。

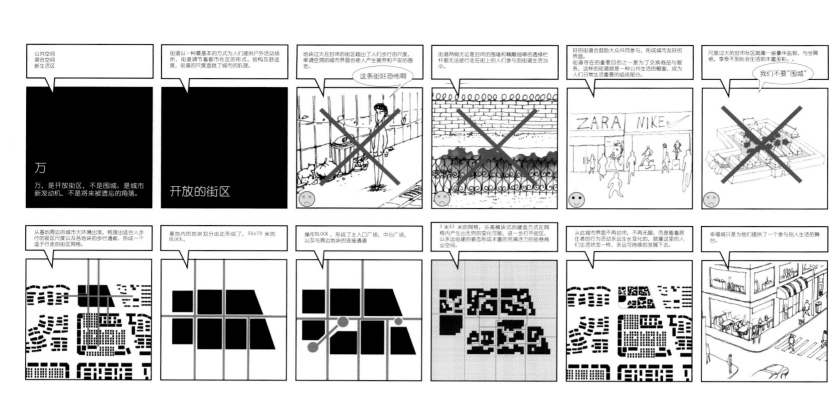

Ground Floor Plan 首层平面图

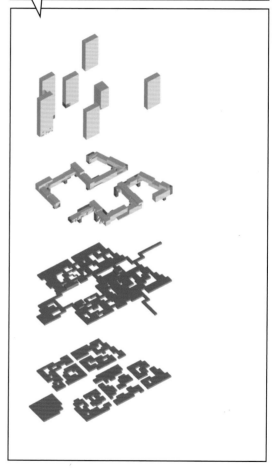

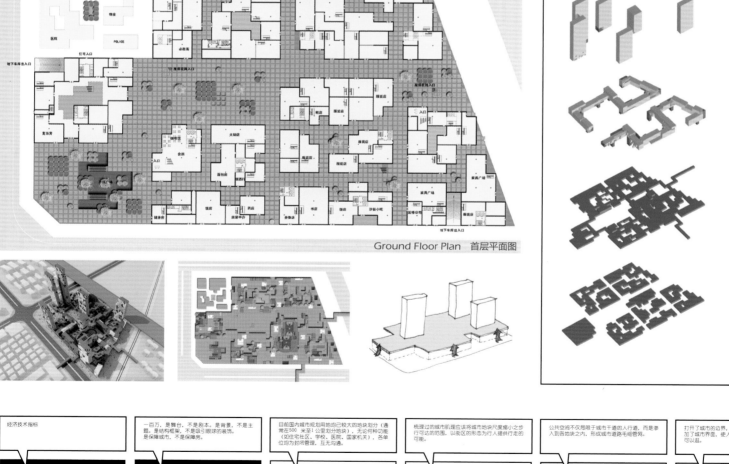

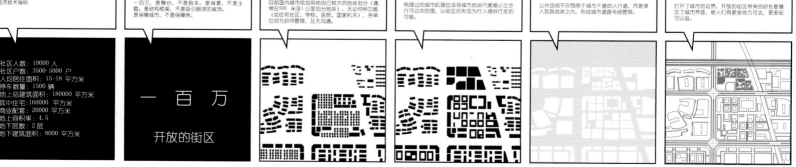

Block Situation Analysis
街区分析图

Section
剖面图

    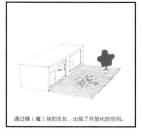 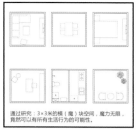

Modulus Design
模数化设计

Modern city faces many problems such as lack of urban space or neighborhood communication. Moreover, the affordable housing challenges become more formidable. The project with an area of 41,694.05 m², expected to accommodate 10,000 people, plot ratio as high as 4.5, will meet the needs of diverse living space and provide life support under the design condition that per capita area is less than 17 m² in a high density, while improving the diversity adaptability of the open space. To this end, the project is named "Happy Town", and proposes three levels and nine concepts from macro to micro, namely, 10,000-public spaces, blending space, organic space; 100-neighborhood spaces, group spaces, landscape garden; 1-private space, family room, home in the sky. In face of construction projects that current affordable housing takes more care on the mass volume and high speed, in the context of standardization, 1 million design of modular construction levels should be made more clear. All kinds of combination of the various possibilities present a mathematical progression growth under the modular modes and are lively in the unity of the rationality. It split the huge residential areas to small recognizable residential blocks, where people can live together as a unit.

现代城市面临着城市空间缺失、邻里交流缺失等诸多问题，而保障性住房挑战更为艰巨。方案用地面积 41 694.05 平方米，预期容纳 1 万人，容积率高达 4.5，在这样一个高密度并且人均面积不足 17 平方米的设计前提下，要满足多样的居住空间需求，并提供完善的生活配套，同时提高开放空间的多样性适应能力。为此，方案命名为"幸福城"。从宏观到微观提出了三个层次、九个概念，即：万——公共空间，混合空间，有机空间；百——邻里空间，组团空间，风景庭院；一——私人空间，家庭空间，宅在空中。面对目前保障房中大批量、高速度的建设项目，在标准化的前提下，一百万模数化建造的设计层次应更为分明，模块化的方式下各种组合的多样可能性呈数学级数增长，在理性的统一下活泼多变。将巨大的居住区分割、划小成为一个可识别的居住街区，人们以居住街区为单位集合居住在一起。

Unit Plan
户型图

住宅采用南北向单廊式，东西向双廊式布局，公共空间面向内庭院。

FEATURE 特点分析

FAÇADE

Uneven balconies break the original stiffness of the façade, increasing a sense of space and three-dimensional effect to the building. At the same time, glass windows not only expand horizons, but also make the building look in a modern appearance.

立面

参差不齐的阳台突破了原有立面的呆板，增加了建筑的空间感和立体感。同时，玻璃窗的运用不仅拓展了视野，也使得建筑看起来更现代化。

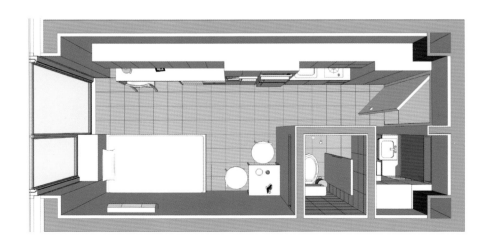

COLOUR 颜色

FAÇADE 立面

SHAPE 造型

WINDOW 窗口

"LA LIBERTE" Housing and Office Building

"自由" 住宅与办公综合楼

Architect: Dominique Perrault Architecture
Location: Groningen, the Netherlands
Photography: Jim Ernst, Prima Focus, Mark Sekuur, DPA, Adagp
Sketches: Dominique Perrault Architecture, Adagp
Renderings and Plans: Dominique Perrault Architecture, Adagp

设计公司：多米尼克·佩罗建筑事务所
地点：荷兰格罗宁根省
摄 影：Jim Ernst 、Prima Focus、Mark Sekuur、DPA、Adagp
草图：多米尼克·佩罗建筑事务所、Adagp
效果图和规划图：多米尼克·佩罗建筑事务所、Adagp

The scheme is a mixed – use building of social housing and offices. The construction of "LA LIBERTE" Housing and Office Building is part of the Ring Zuid Groningen initiative, an agreement of the Eurlings Ministry of Transport, the municipality and province of Groningen to develop the surrounding areas of the "southern ring road of Groningen", the Weg der Verenigde Naties highway.

本案是一栋集合社会住房和办公室的综合建筑。"自由"住宅与办公综合楼是荷兰格罗宁根市环南开发项目的一部分，根据 Eurlings 交通部的协议，在格罗宁根省的各市之间的周边地区开展"格罗宁根南部的二环路发展项目"，Weg der Verenigde Naties 高速路也是其中一部分。

Plan du rez-de-chaussée
Ground floor plan

Site Plan
总平面图

FAÇADE

The façade is divided by volumes and colors: the upper is light while the lower is deep, which gives a stocky and solid sense. The prominent windows increase a three-dimensional effect for the building, which seems to add a dynamic sense.

立面

建筑立面以色彩和体量做了划分，上层颜色较浅，下层较深，给人一种敦实感。另外，突出的窗口增加了建筑的立体感，使建筑看起来更动感。

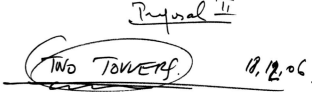

Sketch
手绘草图

The buildings are both made up of a platform, entirely in glass, independent and with the same height (R+2), accommodating the offices. As they are not taller than the nearby blocks, the platforms respect and extend their horizontality.

Then, the two blocks, with different heights, seem to be floating above the platforms and accommodate the housings. Here the architect plays with the volumes: actually the tower A is made up of two volumes of housing with equivalent proportions, and slightly shifted. It seems that the architect has piled up different volumes, one on the others; one volume of offices and one volume of housing for the tower B, one volume of offices and two shifted volumes of housing for the tower A.

The housing blocks are hanging above the offices thanks to a 5-m-high terrace. This "in-between", only a core sheltering the common spaces and brings an easy transition between the private spaces and the offices.

At last, a footbridge, located at the same level and opened to the users, links up the both towers.

两栋建筑都设置有一个平台，全玻璃立面、独立、有相同的高度 (R + 2) 和相适应的办公室。

但平台并不比附近大楼高，而是呈水平方向向外扩展。

两座不同高度的建筑似乎漂浮在平台之上，却又能和周围住宅保持协调。在这里，建筑师与两栋建筑做起了游戏：其实 A 塔楼是由两个等比例又略有变化的住宅体量组成的。看起来就像是将不同的体量堆积起来：B 塔楼有一个办公体量和住宅体量，A 塔楼有一个办公体量和两个变化的住宅体量。

住宅体量悬挂在办公体量 5 米高的地方，"中间"是一个核心，作为私人空间和办公区之间的过渡。

最后，在同一水平位置有一座方便用户通行的人行天桥将两塔相连。

Through the treatment of the façades, Dominique Perrault creates a real dialogue between the two buildings and between the project and its urban environment.

Whatever the viewpoint is, the façades of the three housing blocks never offer the same treatment of colours—black, grey and white. These shades of colour strengthen the stack impression and energize the city skyline.

Moreover, some polished steel panels are placed perpendicularly to the façades, which punctuate the façades and multiply the perceptions of the building.

通过立面的调整，设计师在两座建筑之间以及项目与城市之间创建了一次真正的对话。

无论从哪个视角看，三个住宅体量的立面都没有设置相同的颜色——黑色、灰色和白色。这些颜色的暗部加强了堆叠感，强化了城市天际线。

此外一些抛光钢面板呈垂直方向放置在立面，使得立面得到了强化，加强了公众对建筑的印象。

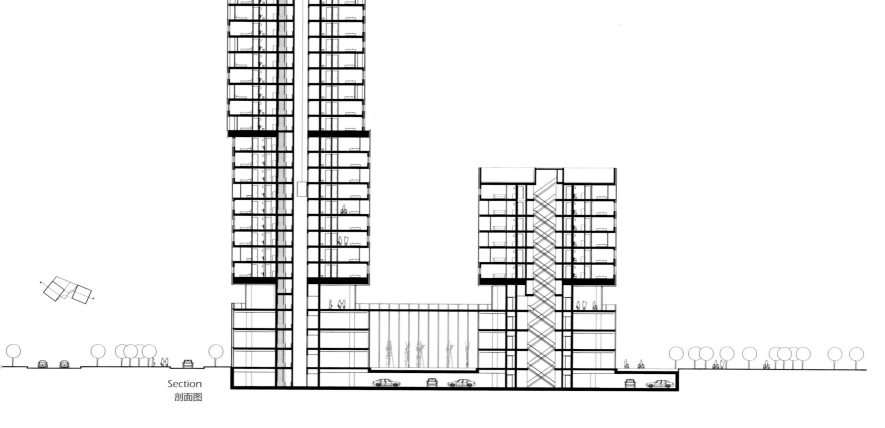

Section
剖面图

MATERIAL 材料

SHAPE 造型

FAÇADE 立面

WINDOW 窗口

Aura

Aura 大厦

Architect: SOMA 设计公司：SOMA 建筑事务所
Location: Erbil, Iraq 地点：伊拉克埃尔比勒市
Photography: SOMA 摄影：SOMA 建筑事务所

The design of the new city centre of Erbil, Iraq aims at creating an icon which would be a main landmark for the city. The design strategy focuses on the extrusion of rectangular shapes from the ground while embracing the land located in front of the buildings. The orientation of the volumes is determined in a way to maximize the sun exposure and natural ventilation, in order to create comfortable spaces for the residents. The landscape is designed according to flexible modules allowing for several activities from swimming, exercising, to meditation. The master plan includes retail, offices, and residential units.

本案作为伊拉克埃尔比勒市的新城市中心，旨在打造一个全新的城市地标。设计方案重点在于突出拔地而起的长方形体量，并环抱建筑前方的土地。体量的朝向考虑到了阳光辐射和自然通风，以为居民提供一个舒适的空间。景观设计是根据灵活的模块空间进行的，可以进行游泳、锻炼和冥想等活动。总体规划还包括了商场、办公室和住宅单位等。

FEATURE 特点分析

SHAPE

The modeling of buildings with different heights is elegant and solemn. Bathed in daylight, rested in the sunset, it seems brilliant with the cooperation of light, and echoes with the surrounding landscapes.

造型

建筑高低错落的造型大气又典雅，在日光里沐浴，在夕阳中小憩，在光亮的配合下褶褶生辉，并同周边景观形成良好的辉映。

SUSTAINABILITY 可持续性

FAÇADE 立面

STRUCTURE 结构

SHAPE 造型

YDA

YDA 公寓大楼

Architect: Gokhan Avcioglu & GAD	设计公司：Gokhan Avcioglu & GAD
Client: YDA	客户：YDA
Location: Istanbul, Turkey	地点：土耳其伊斯坦布尔市
Site Area: 2, 500 m²	占地面积：2 500 平方米
Gross Floor Area: 45,000 m²	建筑面积：45 000 平方米
Floors: 20	层数：20
Height: 75.5 m	高度：75.5 米
Realization: 2013	完成时间：2013 年

YDA is a residential project located in Bakırköy, Istanbul. The project site, near the Marmara Sea, is located close to a formerly industrial area Veliefendi. The place was once home to Istanbul's leather tanneries. This area has undergone a significant transformation and around late 19th century it became one of the most desirable residential exurbs of the city. Today Bakırköy is not only popular with its huge shopping district, but also it is a really large housing area.

项目坐落于伊斯坦布尔的巴克科伊，濒临希腊马尔马拉海，临近 Veliefendi 前工业区。这个地方曾经是伊斯坦布尔皮革厂所在地，并经历了一次重大的变革。到 19 世纪晚期，这里成为城市最理想的居住郊区之一。如今，巴克科伊并不是一个仅仅依靠大型购物区而闻名的地区，它同样是一个庞大的居住地区。

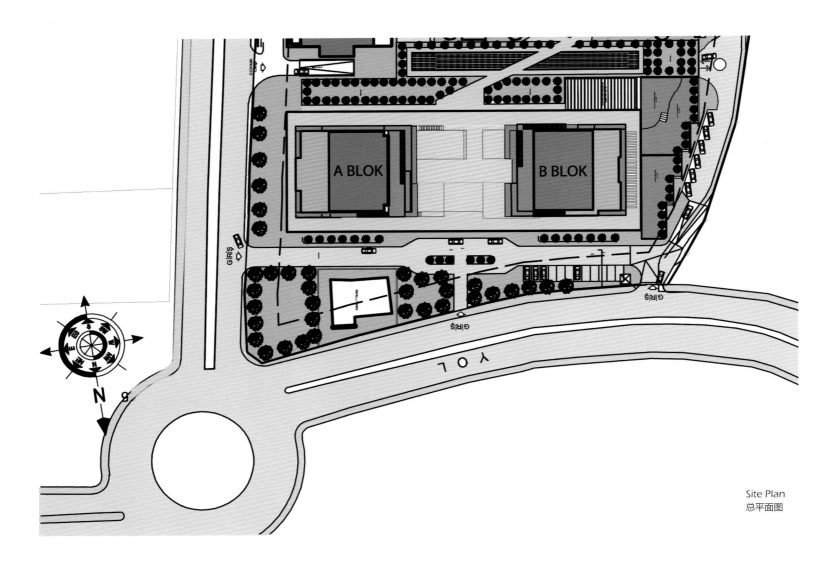

Site Plan
总平面图

While the site is not directly adjacent to the sea, view corridors from the proposed tower are unobstructed by development, which afford 360-degree vistas of Istanbul and the Marmara Sea coast. The potential for these view corridors drives the concept for YDA.

Placed on the east-west axis, YDA consists of two residential blocks on a shared plinth. The building program includes residences, retail space, restaurants and social activities. The base consists of two levels for retail and common activities of residents, while there are tables and chairs located in the public space in the shared plinth, used for leisure breaks, and allowing people to enjoy the surrounding scenery when they are in the relaxation. So this human design must be liked by people.

这个地区不是直接邻海，但是通过建设发展，站在建筑中也可以尽览周边所有环境，并且可以 360 度全方位地欣赏来自伊斯坦布尔和马尔马拉海岸的绝美风景。正是这种潜在的美景驱动了 YDA 项目的开发。

构成 YDA 公寓大楼的两个住宅区被安置在同一个裙楼上，呈东西方向设置。建筑项目包括住宅区、零售区、餐饮区和社会活动区。这个裙楼高两层，主要用于零售和居民的公共活动。而位于裙楼上的公共空间设置有桌椅、凉亭，供人们闲暇休息之用，让人们在放松休息的时候还可以欣赏周边的风景。这样极富人情味的设计必定受到人们的喜爱。

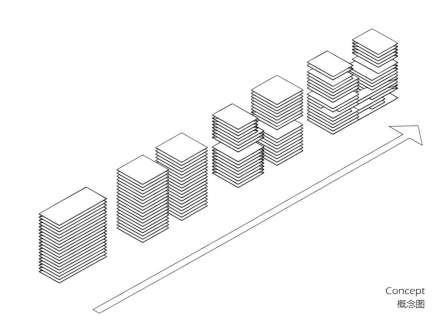

Concept
概念图

As the building rises, the two towers move away from each other to open the façades for panoramic views and to obtain daylight. The façade is a combination of continuous glass and extending terrace slabs which differ according to the size of the apartments.

随着建筑高度的增加，塔楼间距也越来越远，以获得更广阔的全景视野和更充足的日光。建筑立面由连续玻璃和扩展的阳台石板组成，这些阳台石板根据公寓的不同规格而不同。

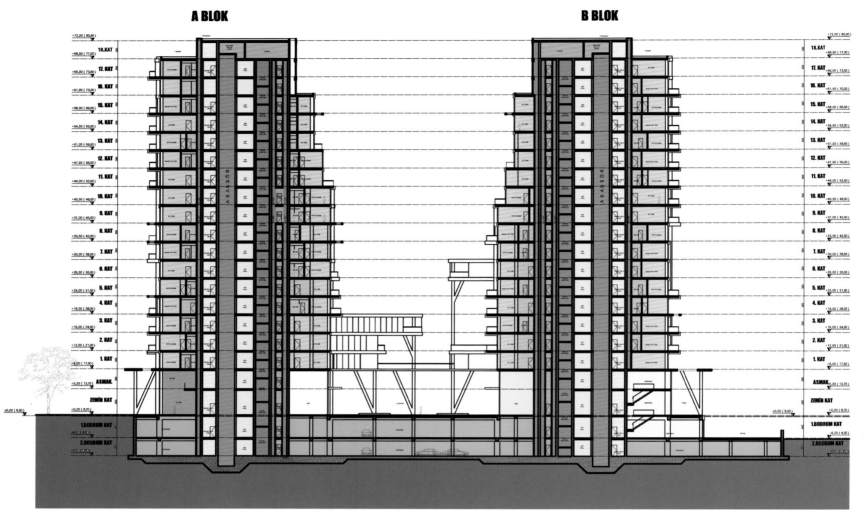

A-A Section
A-A 剖面图

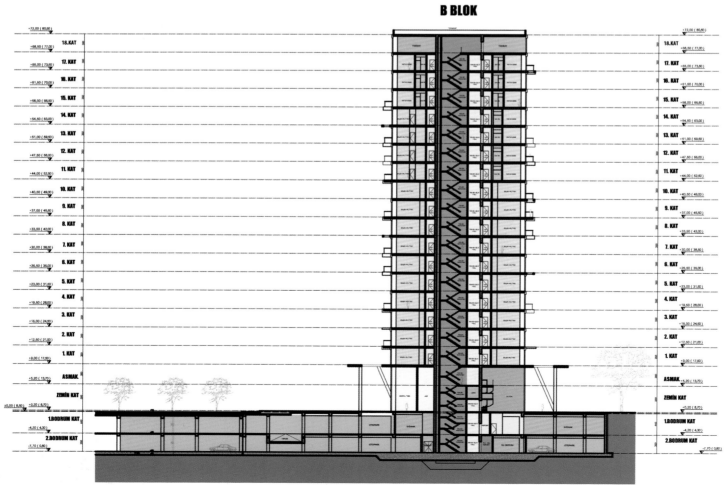

C-C Section
C-C 剖面图

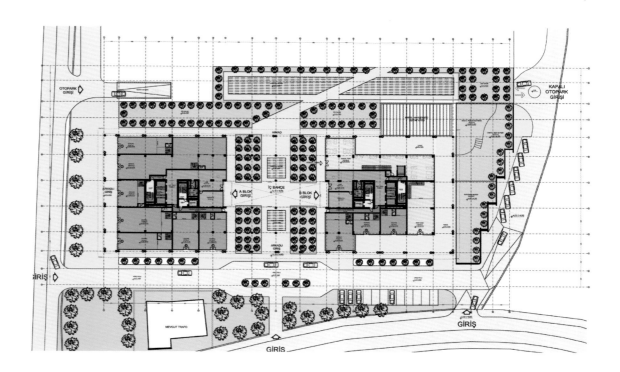

Ground Floor Plan
首层平面图

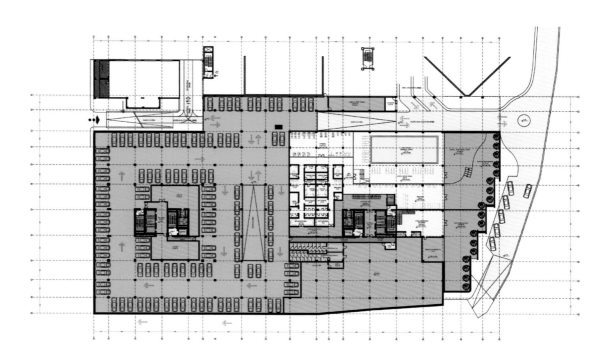

Floor Plan
楼层平面图

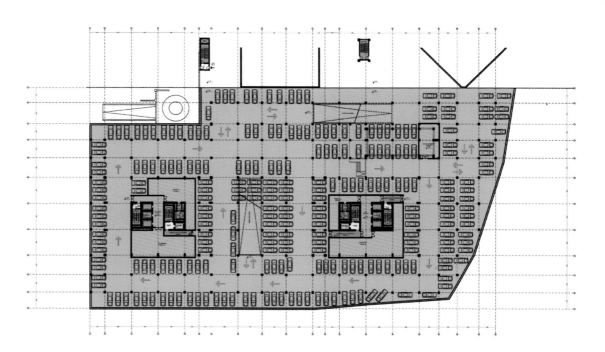

Floor Plan
楼层平面图

FEATURE 特点分析

FAÇADE

Green vegetation embedded in the façade added a dash of bright colour to the icy tones, let the concrete buildings come back to life, made a perfect integration of architecture and ecology and conveyed the architectures ecological design concept.

立面

绿色植被被嵌入立面中，为原本冰冷的色调增添了一抹亮色，又让水泥建筑活了起来，真正地做到了建筑与生态的完美结合，传达了建筑的生态设计理念。

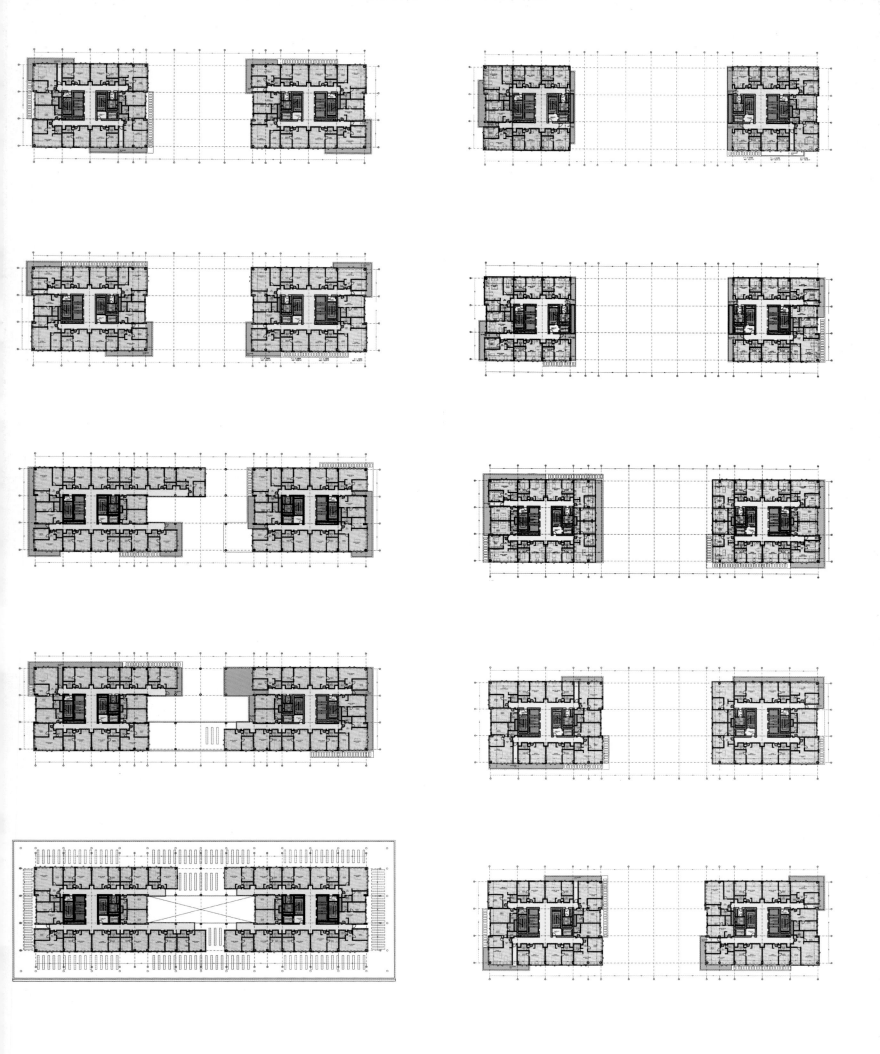

Floor Plan
楼层图

MATERIAL 材料

SHAPE 造型

STRUCTURE 结构

FAÇADE 立面

3 Midtown, Miami, USA

美国迈阿密 3 号中心城大厦

Architect: Oppenheim Architecture + Design
Location: Miami, Florida, USA
Site Area: 74,322.43 m²

设计公司：奥本海姆建筑设计公司
地点：美国佛罗里达州迈阿密市
占地面积：74 322.43 平方米

3 Midtown is a microcosm of the city and metropolitan life. Situated on an entire block in Midtown Miami, this sizable project is broken down into smaller components allowing for a reduction in perceived mass and inherent anonymity. The architectural infrastructure, compliant with the urban plan, creates a harmonic diversity of unique experiences and living environments (nearly 40 unit varieties).

3 号中心城大厦是城市和都市生活的一个缩影。项目整个坐落于迈阿密市中心区，大的规模被分解成诸多小的领域以减少感官上庞大的体量和固有的无名性。建筑的基础设施符合城市规划，创建了一种多样性的独特经历和生活环境的和谐（近 40 种不同的单元）。

Location
区位图

The design concentrates the bustling metropolises life in a small space through a careful layout that transforms the simple block into a bustling space. The unique shape makes the building look different, but intimately link with the surrounding environment, and together play on a harmonious concerto with the city.

In addition to having multiple lobbies at street level, approximately 80% of the pedestrian experience is commercial-based activities at various scales—from the intimate mews shops to the more spacious tenant opportunities. All parking and other service-related zones of the project are properly concealed by active program of a mixed-use nature. Shifting south and west at the second level, the massing of the block creates a protected promenade of sensible proportion for people to enjoy the street-side shops.

项目通过精心的布局，将繁华的大都市生活浓缩在小型空间里，将原本简单的街区打造成为繁华独特的另类空间。独特的造型设计让建筑看起来与众不同，却与周边环境紧密相连，同城市演绎一段和谐的协奏曲。

除了在首层拥有多功能大厅外，约 80% 的行人都会徘徊在不同规模的商业区之间——从私人商店到大型商户，应有尽有。所有的停车和其他相关服务区域通过活动区域的多功能性质被巧妙地隐藏起来。在建筑的二层将南部和西部对换，这样，建筑的体量便形成了比例适当的荫凉散步长廊，以供人们在街边小店之间游荡。

FEATURE 特点分析

SHAPE

This appealing urban collage mutates both vertically and horizontally around the mostly glazed block, signifying divergences in residential types while responding to contextual relationships through variations in porosity. As a multi-purpose device, this pattern sensibly reacts to varying degrees of privacy in the tight urban fabric proposed while creating porches, verandas, loggias and other spatially sublime moments.

造型

这个吸引人的城市布局从垂直和水平两个方向上围绕着大面积釉面玻璃区进行调整，既表现了住宅类型的差异同时又以其多孔的另类住宅类型和周围的环境相呼应。作为一种多用途策略，这种模式对紧凑城市结构中的不同隐私程度做出敏感反应，提供了门廊、阳台、凉廊和其他别样的空间。

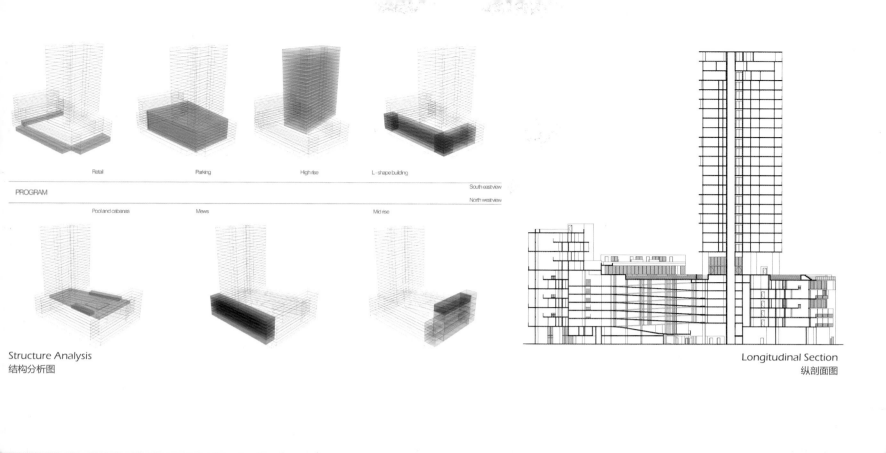

Retail Parking High rise L-shape building

PROGRAM

South east view

North west view

Pool and cabanas Mews Mid rise

Structure Analysis
结构分析图

Longitudinal Section
纵剖面图

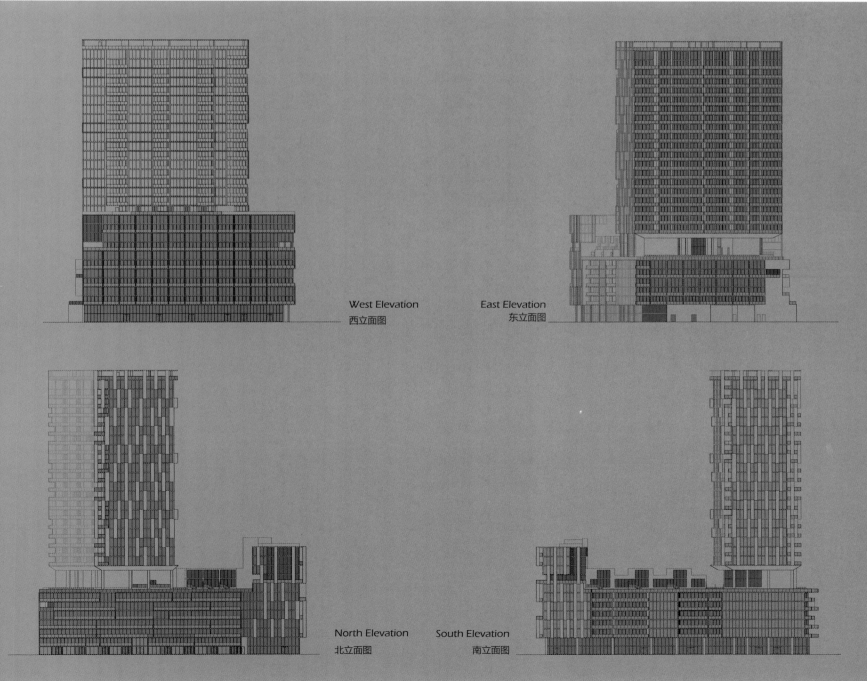

West Elevation
西立面图

East Elevation
东立面图

North Elevation
北立面图

South Elevation
南立面图

Set on a diagonal above a communal sky garden, the tower component initiates an urban dance, gently twisting off axis and out of the way of its neighbors, establishing optimum view corridors and separations. A dynamic and systematic rhythm of solid and void abstracts the typological reference while paradoxically generating allusions of urban ensembles.

This intentional distortion creates a building (or city block) that is simultaneously novel sculpture and familiar edifice—evoking notions of beautiful and charming villages where a monochromatic fabric of similar form and element drapes over masses of buildings to provide spaces for life's enjoyment.

建筑位于对角线处一座公共空中花园上方，各部分相互配合上演了一段城市舞蹈，轻轻地扭动着直到脱离了群体，并建立了最佳视觉效果的长廊和距离。虚与实的动态体系节奏将建筑形式抽象地表现了出来，同时又有些矛盾地与整个城市产生了合奏。

这种有意的变形创建了如此的建筑（或街区），就像是新的雕塑品或似曾相识的建筑物——唤起人们对美丽迷人村庄的向往，那里造型相似的单色建筑和大量建筑带来的元素为生活娱乐提供了空间。

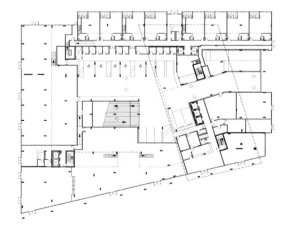

Ground Floor Plan
首层平面图

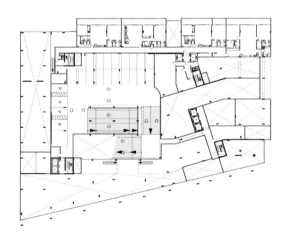

Second Floor Plan
2层平面图

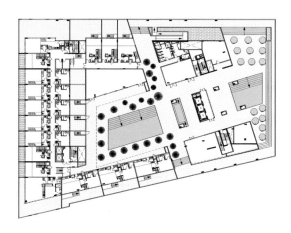

Pool Deck Floor Plan
泳池楼层平面图

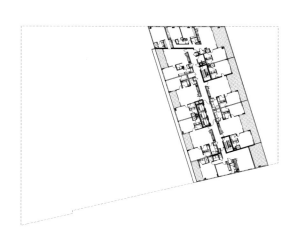

Typical Tower Floor Plan
塔楼平面图

VIEW 视野

SHAPE 造型

MATERIAL 材料

FAÇADE 立面

Continental Tower, Stockholm, Sweden

瑞典斯德哥尔摩大陆塔

Architect: C. F. Møller Architects
Location: Stockholm, Sweden
Gross Floor Area: 20,000 m²
Floors: 30
Height: 100 m
Renderings: C. F. Møller Architects

设计公司：C. F. Møller Architects
地点：瑞典斯德哥尔摩市
建筑面积：20 000 平方米
层数：30
高度：100 米
效果图：C. F. Møller Architects

The tower block construction in Stockholm has been planned in connection with the extension of the City Line rail link.

大陆塔项目坐落于瑞典的斯德哥尔摩市，规划旨在连接并拓展城市铁路线。

Location Plan
区位图

The complex of 100 m, the equivalent of approximately 30 storeys, will include hotels, offices and possibly housing.

The building will be equipped with an entrance to a new station at street level directly opposite the city's old main railway station, Centralen, and will therefore be a landmark for Stockholm's new metro.

C. F. Møller's design is based on a concept of multi-function, a building where many varying uses can take place simultaneously.

The tower draws inspiration from the city of Stockholm, an example of which is Stockholm's largest church Klara kyrka situated right next to the site.

The project reinterprets the city's beautiful town hall and classic church steeples, which combinedly makes up the city's historic landmarks and creates Stockholm's historic skyline, but at the same time integrates the unique urban feeling defining the area around Centralstationen.

这个综合建筑高达 100 米，相当于 30 层楼高，包括酒店、办公和潜在居住空间。

建筑将会配备一个直通街道新车站的入口，那个新车站与城市旧车站 Centralen 刚好对立设置，项目会因此成为斯德哥尔摩地铁的里程碑式建筑。

C. F. Møller 的项目设计以多功能概念为基础，一栋建筑将允许多种功能设施同时运作。

建筑的设计灵感来源于斯德哥尔摩城市，斯德哥尔摩最大的教堂 Klara kyrka 就位于其右边。

该项目将城市美丽的市政厅和经典的教堂尖顶做了全新的解释，将这些组合起来成为城市的历史地标，并创建了斯德哥尔摩的历史性的天际线，但同时融合了独特的城市感觉并对围绕着中央火车站的该区做了新的定义。

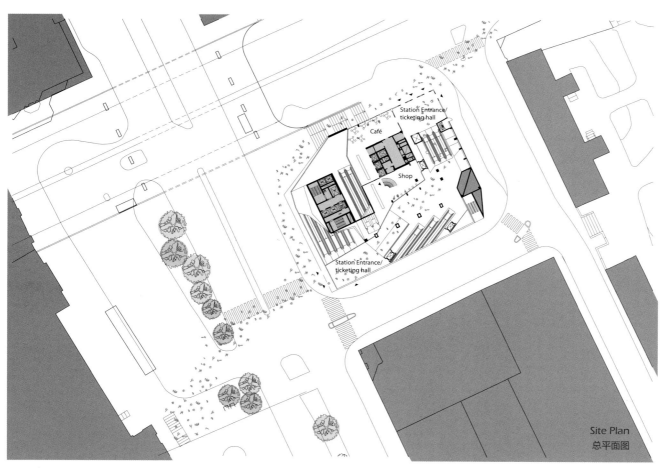

Site Plan
总平面图

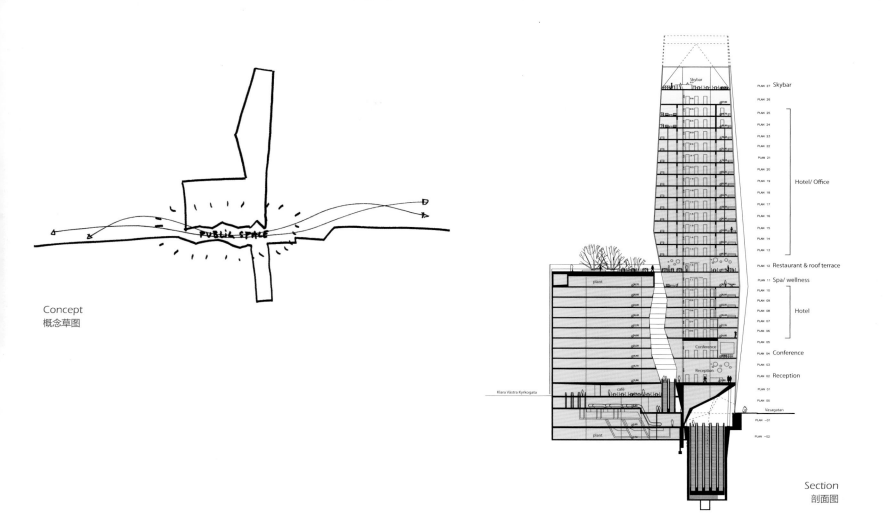

Concept
概念草图

Section
剖面图

Within the section diagram:

Skybar
PLAN 27 Skybar
PLAN 26
PLAN 25
PLAN 24
PLAN 23
PLAN 22
PLAN 21
PLAN 20   Hotel/ Office
PLAN 19
PLAN 18
PLAN 17
PLAN 16
PLAN 15
PLAN 14
PLAN 13
PLAN 12 Restaurant & roof terrace
PLAN 11 Spa/ wellness
PLAN 10
PLAN 09
PLAN 08   Hotel
PLAN 07
PLAN 06
PLAN 05
PLAN 04 Conference
PLAN 03
PLAN 02 Reception
PLAN 01
PLAN 00
Vasagatan
PLAN −01
PLAN −02

plant
café
Klara Västra Kyrkogata
Conference
Reception
plant

SHAPE

The tower draws inspiration from the city of Stockholm, while the cutting stripe also makes the building become an indispensable landscape. Irregular stereoscopic design matches with dazzling tonal shape, makes the building seem luxurious.

造型

建筑的设计灵感来源于斯德哥尔摩城市，其切割式的造型也让建筑成为了一道不可或缺的风景。不规则立体设计配合耀眼夺目的色调，让建筑看起来奢华无比。

East Elevation
东立面

South Elevation
南立面

West Elevation
西立面

North Elevation
北立面

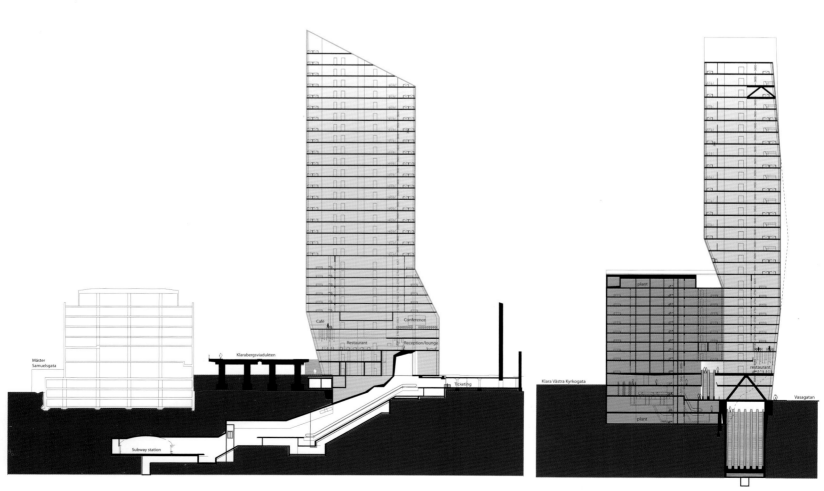

Mäster
Samuelsgata

Klarabergsviadukten

Café

Conference

Restaurant

Reception/lounge

Ticketing

Subway station

plant

Klara Västra Kyrkogata

plant

restaurant

Vasagatan

Section(Subway)
剖面图（地铁）

Section(Escalator)
剖面图（电梯）

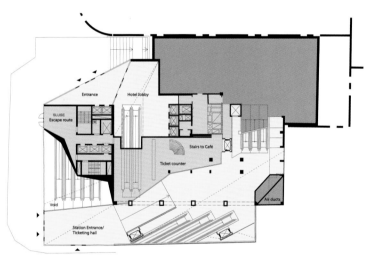

3rd Floor Plan - Lower-Groundfloor Entrance
3 层平面图 - 底层下部入口

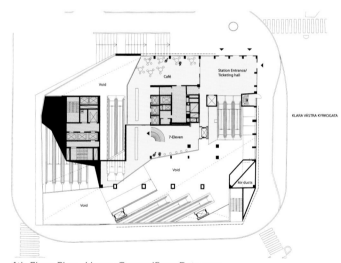

4th Floor Plan - Upper Groundfloor Entrance
4 层平面图 - 底层上部入口

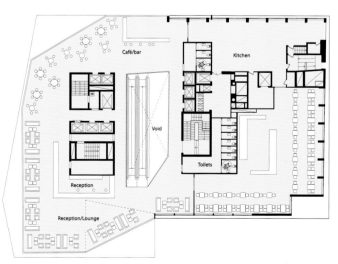

5th Floor Plan - Reception
5 层平面图 - 接待区

6th Floor Plan - Conference Centre
6 层平面图 - 会议中心

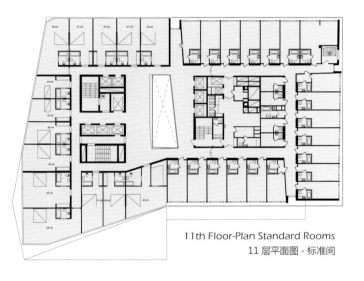

11th Floor-Plan Standard Rooms
11 层平面图 - 标准间

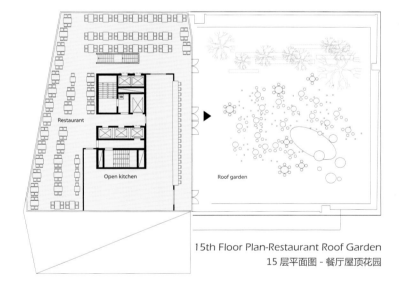

15th Floor Plan-Restaurant Roof Garden
15 层平面图 - 餐厅屋顶花园

16th Floor Plan - Deluxe Rooms
16 层平面图 - 复式公寓

17th Floor Plan - Luxury Apartment
17 层平面图 - 豪华公寓

FAÇADE 立面

SHAPE 造型

VIEW 视野

GREENERY 绿色

Beach and Howe St, Vancouver, Canada

加拿大温哥华滨海大道与豪威街交界处的滨豪大厦

Architect: BIG, Westbank, Dialog, Cobalt, PFS, Buro Happold, Glotman Simpson, James Cheng
Client: Westbank Projects Corp
Location: Vancouver, Canada
Gross Floor Area: 60,670 m²
Floors: 49
Height: 150 m
Renderings: Luxigon, Glessner, BIG
Realization: 2012

设计公司：BIG、Westbank、Dialog、Cobalt、PFS、Buro Happold、Glotman Simpson、James Cheng
客户：Westbank Projects Corp.
地点：加拿大温哥华市
建筑面积：60 670 平方米
层数：49
高度：150 米
效果图：Luxigon、Glessner、BIG
完成时间：2012 年

The 149.35-m-tall Beach and Howe mixed-use tower by BIG, Westbank, Dialog, Cobalt, PFS, Buro Happold, Glotman Simpson and local architect James Cheng marks the entry point to downtown Vancouver, forming a welcoming gateway to the city, while adding another unique structure to the Vancouver skyline.

高 149.3 米的 Beach and Howe 多功能塔楼由 BIG、Dialog、Cobalt、PFS、Buro Happol、Glotman Simpson 和地方建筑师 James Cheng 共同建造，是温哥华市区的入口标志，形成了一座欢迎人们进入城区的大门，也为温哥华的天际线增添了一座独特的建筑物。

Site Analysis
场地分析图

SITE

STREET SETBACKS TO ALLOW FUTURE WIDENING OF HOWE STREET AND PACIFIC STREET

5 METER SETBACKS FROM GRANVILLE BRIDGE EDGE, TRISECTING THE SITE

SAFETY SETBACK OF 30 METERS FROM GRANVILLE BRIDGE

RESULTING TRIANGULAR FOOTPRINT OF APROX. 6,000 SF

Construction Analysis
建设分析图

EXTRUSION OF RESULTING FOOTPRINT

RESULTANT SILHOUETTE RESEMBLES A CURTAIN BEING DRAWN ASIDE, WELCOMING PEOPLE AS THEY TRAVEL ACROSS THE GRANVILLE BRIDGE

30 OPPORTUNITY AWAY FROM THE BRIDGE, THE TOWER CAN RECLAIM ITS LOGICAL FOOTPRINT AT THE TOP

CREATION OF CONTINUOUS ACCESSIBLE PLATFORMS FROM PACIFIC ST

RAISING THREE NEW PEAKS TO EMPHASIZE FLOWS ALONG GRANVILLE BRIDGE

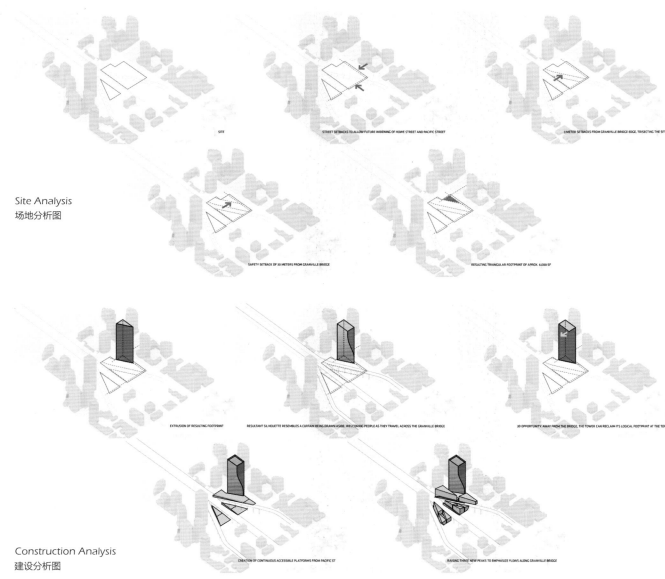

BIG's proposal, named after its location on the corner of Beach and Howe Street next to the Granville Street Bridge in downtown Vancouver, calls for 600 residential units occupying the 49-storey tower, which would become the city's fourth tallest buildings. The tower is situated on a 9-storey podium base offering market-rental housing with a mix of commercial and retail space.

The tower takes its shape after the site's complex urban conditions aiming to optimize the conditions for its future inhabitants in the air as well as the street level. As its base, the footprint of the tower is conditioned by concerns for two significant neighboring elements, including a 30-m setback from the Granville Bridge which ensures that no resident will have windows and balconies in the middle of heavy traffic as well as concerns for sunlight to an adjacent park which limits how far south the building can be constructed. As a result, the footprint is restricted to a small triangle.

项目由 BIG 提出，坐落在温哥华市中心区毗邻格兰维尔街大桥的滨海大道与豪威街交界处角，因此得名"滨豪大厦"。它有 600 套住宅，49 层高，将成为城市的第四高建筑。高楼坐落在一个 9 层的基座上，这里提供商业和零售混合功能的可出租房屋。

高楼建成后，这一地块的城市改造综合目标是为未来的居民改善空气条件，改善街道环境。作为它的基础，高楼的底部所涉及的条件更因两个重要的邻近居住单元而受关注，包括从格兰维尔桥后退 30 米，以确保没有居民的窗子或阳台会处在交通拥堵中间，同时还要关注邻近公园的日照，它限定了其南面的高楼必须相距多远才能建造，由此决定了高楼的底部范围被限定成一个很小的三角形地块。

As the tower ascends, it clears the noise, exhaust, and visual invasion of the Granville Bridge. BIG's design reclaims the lost area as the tower clears the zone of influence of the bridge, gradually cantilevering over the site. This movement turns the inefficient triangle into an optimal rectangular floor plate, increasing the desirable spaces for living at its top, while freeing up a generous public space at its base. The resultant silhouette has a unique appearance that changes from every angle and resembles a curtain being drawn aside, welcoming people as they enter the city from the bridge.

The courtyards created by the building volumes, roofs and terraces are all designed to enhance views from the Granville Bridge and the residential units above. The canted, triangular clusters of green roofs create a highly graphic and iconic gateway to and from the downtown core, reinforcing the City of Vancouver's focus on sustainable cities. The exterior façades respond to the various solar exposures, which is integral to the overall sustainability concept. The building will strive for LEED Gold Certification.

随着高楼的上升，格兰维尔大桥的噪声、废气和视觉污染也清楚地呈现。BIG 的设计就是改造这些嘈杂的地区，就像在高楼上清楚地看到的受大桥影响的区域一样。这些改造活动把无效的三角形空间变成了最佳的矩形平面，为顶上生活增添了令人满意的空间，同时又腾出了底部的大量公共空间。高楼独特显现出的轮廓，从各个角度看都有不同变化，像一条从旁边垂下的幕布，欢迎着从大桥进入城市的人们。

由建筑所形成的院子、屋顶和阶地都经过设计，以增加从格兰维尔桥和上面居住单元里看上去时的景观。倾斜的三角形绿色屋顶形成了非常形象的进出市中心的通行图标，增强了温哥华城作为可持续发展城市的聚焦度。高楼的外立面多方面反映光照，充分折射出可持续发展的观念。项目将力求获 LEED 金奖。

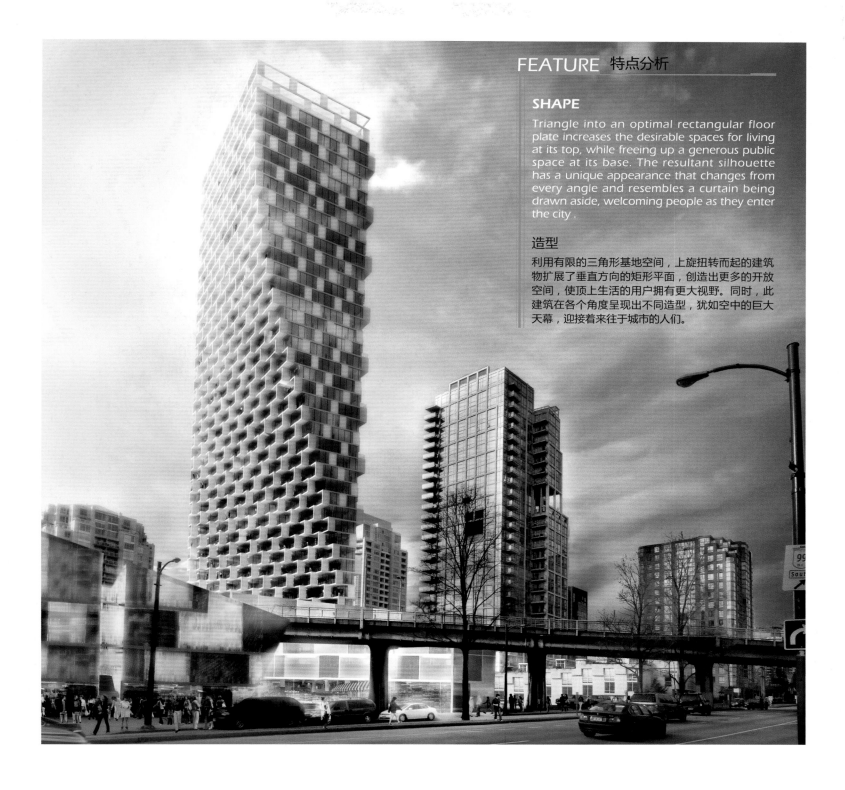

FEATURE 特点分析

SHAPE

Triangle into an optimal rectangular floor plate increases the desirable spaces for living at its top, while freeing up a generous public space at its base. The resultant silhouette has a unique appearance that changes from every angle and resembles a curtain being drawn aside, welcoming people as they enter the city .

造型

利用有限的三角形基地空间，上旋扭转而起的建筑物扩展了垂直方向的矩形平面，创造出更多的开放空间，使顶上生活的用户拥有更大视野。同时，此建筑在各个角度呈现出不同造型，犹如空中的巨大天幕，迎接着来往于城市的人们。

FAÇADE 立面

ECOLOGY 生态

STRUCTURE 结构

ENERGY SAVING 节能

Essentials Dream Town Qingdao – Amsterdam Block

青岛梦想之城 - 阿姆斯特丹街区

Architect: Contexture Architects
Client: China Vanke Qingdao
Location: Qingdao, China
Retail Area: 5,000 m²

设计公司：Contexture Architects
客户：青岛万科集团
地点：中国青岛市
零售区面积：5 000 平方米

Here on plot A6, there has been created the perfect combination of the liveliness of the city with its nice small boutiques and terraces on the one hand, and the relaxation of the national park with its beautiful trees and immense green area for strolling on the other hand. Therefore, two urban blocks have been formed around green courtyards which are attached to the mountain park.

A6 地块设计将城市生活和自然景观完美地结合在一起，前拥精致的商业街和大量的休闲平台，背靠双山自然公园，绿意盎然。因此两个城市公寓各自环绕景观庭院，将双山的景观引入居住社区。

Instead of making apartment towers as solitary objects, we try to make enjoyable urban spaces: streets, squares and courtyards. On two sides of the green central axis square which leads to the mountain park, low apartment blocks have been situated which form an intimate square with terraces belonging to the coffee shops and small restaurants which are located on the ground floor. Near the Hefei Road the block has been set back and thus also creates a large square which ascends slowly and has various levels.

Everywhere in China in new housing development projects, one can see the repetition of identical units both vertically and horizontally. There is much uniformity. Here, in plot A6, a new strategy has been followed: the blocks have been brought back into hierarchy and the houses have been individualized. This has been done in order to bring back human scale in the living environment of the big city life and to make an agreeable and understandable place to live. By distinguishing three levels of perception, the large buildings become an acceptable and even tactile scale.

The first, largest scale level is that of the city, here we can experience individual towers, as identities, from large distance. So, from a distance a range of dissimilar towers can be seen on the highest level of the plot. These "identities" protrude from the building ensemble and are materialized in 3 different kinds of grey plasterwork, from light to dark.

Then , secondly, there is the scale level which still has contact with people on the ground floor, with the storeys up to the eighth or ninth floor. This is in fact the height where one becomes detached from earth activities. This middle level is also made visible in the design by materialization which reminds us of the natural stone in the historical German architecture in Qingdao.

The last scale level is the one on the ground and first floor, which consists of the commercial spaces. This is the place where one can actually touch the building and has the closest perception of things. Therefore materials here are fine and small and there is much detail in the façade.

为了避免公寓楼成为一个个死板的单体，我们试图创造出宜人的城市空间：步行通道、休闲平台、绿化庭院等。在通向双山公园的景观轴两侧公寓楼的高度降低，底层设置咖啡厅、小餐馆等，形成亲近的平台空间。同时，建筑沿着肥路后退，创造出大面积沿街休闲平台，并顺着道路逐级上升，给路人多层次的空间感受。

我们在国内到处可见竖向横向单调地不断重复同一元素的住宅设计。在 A6 地块我们打破了这种常用的设计方式。每栋楼有了自己的特征和层次。这是为了使大城市的生活环境回归人性化的尺度和创造一个宜居的空间。通过对人感知三个层面的区分，将大体量的建筑还原到适宜可触的尺度。

第一个层面，也是最大的层面是城市层面，是从大角度远距离地感受这些建筑。此时隔开一段距离，一排不同的大厦可以在最高点被看到。个性化元素突出建筑立面，并被区分为 3 种不同灰度的墙面，从浅到深。

第二个层面，是地面行人可感知的尺度，大约八、九层楼高。超过这个高度的事物就已经脱离可感知尺度了。这个中间层面的设计从颜色到形式均模仿青岛传统德式建筑中常用的天然石材，形成具有当地传统特色的城市界面。

第三个层面是地面层，以商业空间为主。这个尺度使人们可以近距离地触摸和观察，因此立面的材质及细部设计更加精致。

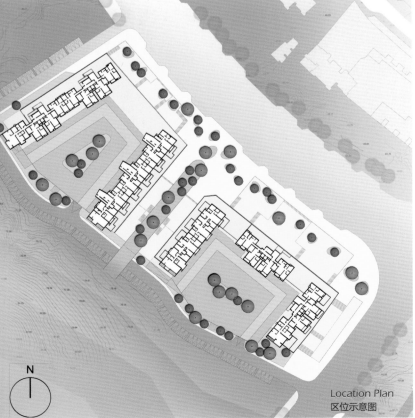

N

Location Plan
区位示意图

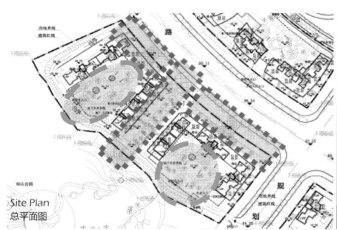

Site Plan
总平面图

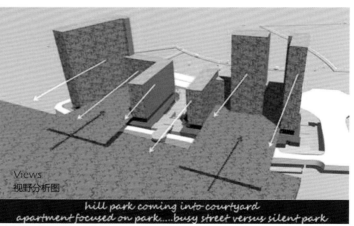

Views
视野分析图

hill park coming into courtyard
apartment focused on park.....busy street versus silent park

Landscape Design
景观设计图

临近双山的四栋公寓楼端户为 160 平方米豪华观景大户型，成为整个 A6 地块最奢华的户型。起居室和餐厅融为一体，使之完全向山体方向开敞并在空间较长一侧采用大面积的玻璃，尽览绿色美景。

此户型由三个卧室，一个书房，两个卫浴室和两个阳台组成。

充满活力多样化是城市的代表特征，而乡村生活更趋于平静和谐。在 A6 地块中我们将两种不同的体验融合在了一起。抹灰面、面砖、铝合金、玻璃为主要选用材料。通过多种材料、不同颜色的运用、细部抽象处理等手法，创造出不同的体块，使每户都具有可识别性，而不是淹没在建筑大体量之中。但同时，我们仍在多样化设计中寻求谐调和秩序感，创造出丰富的表现和清晰简单的形式及体量之间的平衡。

内部庭院的景观设计是根据与街道和城市空间形成的对照而来。我们的设计本着创造原生态有机的自然景观，弱化高楼对环境的影响，让人感觉仍然置身于山景之中。因此，庭院中设计了较高的树篱，葱郁的大树和原生态的草丛。绿色有机的景观和实体建筑形成对比。

On the southwestern edge of the plot, large apartments of 160 m² have been created. These special ones have complete focus on the mountain park and are in fact the most luxurious available units. Their living area and dining room make one big space which is completely related with the park, and on the long side of the space they have glass walls which gives fantastic views on the greenery.

Next to this they consist of three sleeping rooms, one study room, two bathrooms and two terraces.

City feeling is about vitality and dynamics while country side can be best associated with calmness and unity. In the A6 project we have expressed both and made different forms by using only a few materials and colors and by making the details as abstract as possible. Plasterwork, tiles, aluminum and glass are the main materials we choose to achieve this. As noted before we find it important that the resident can simply distinguish his or her house and can easily point out where one settles. Creating one's own place to live instead of being part of an anonymous mega structure is the message. Still it remains to be our task to create order and context in the built environment, therefore we have sought a delicate balance between richness of expression and clearness of simple forms and volumes.

The landscape design for the inner courtyards is based upon the contrast with the street and the urban space. Our idea is to make an organic landscape in the courtyard where one does not experience the high buildings but presumes to be in the mountain park. Therefore high hedges have been planned together with covering trees and wild grasses, as green and informal as they can to be in contrast with the straight forms of the stone buildings.

FEATURE 特点分析

ECOLOGY

The mountain park, the city natural oxygen bar, which is the greatest quality of this location "rolls" down to the Hefei Road; the green courtyards with wild grasses are the extension of the park while even on roadside above the commercial spaces, trees are planted. The greenery extends to every corner of the site.

生态

被称为城市天然氧吧的双山公园是该地块最特殊的资源，设计将双山绿色引入到小区庭院并一直延续到合肥路。含天然草坪的绿色庭院是双山公园的延伸，沿合肥路商业店铺屋顶设计了空中绿化，将绿色渗透到每个角落。

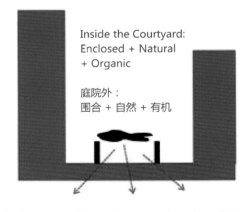

Outside the Courtyard:
Open + Urban

庭院外：
开放 + 城市化

Inside the Courtyard:
Enclosed + Natural
+ Organic

庭院外：
围合 + 自然 + 有机

Outside the Courtyard:
Open + Urban

庭院外：
开放 + 城市化

Minimal experience of blocks around, Orientation on park and sunshine:
Sitting on benches, Under trees, Behind high hedge, Surrounded by wild grasses.

尽量减少视线阻挡，朝向双山公园和日照方向：
在原生态的草丛中，人们可以坐在长凳子上，绿树下，高树篱后休息。

Concept Landscape Courtyard
庭院景观设计概念

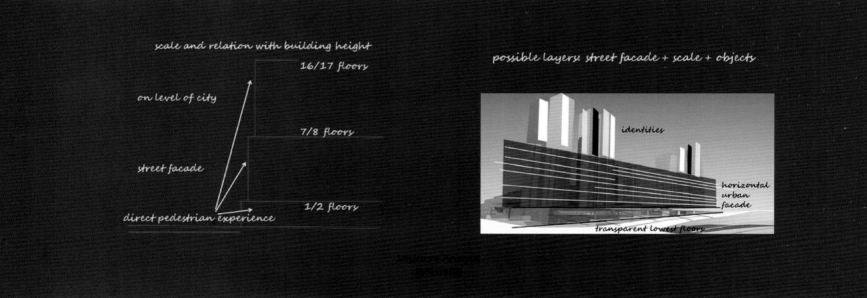

scale and relation with building height

16/17 floors

on level of city

7/8 floors

street facade

direct pedestrian experience

1/2 floors

possible layers: street facade + scale + objects

identities

horizontal urban facade

transparent lowest floors

All apartments have been designed in order to have the optimal effective use of the space. And at the same time, every effort has been made to have a spatial relation between the different rooms which enlarges the apartment.

In block 7, 10 and 11 on each of the sixteen floors we find four apartments of around 90 m² being accessed from the elevator hall. They all have a comfortable living room which is in open connection with the dining area. The closed kitchen is situated right next to this. The apartments consists of two bedrooms and one or two terraces, while the kitchen and bathroom of course can open to outside for fresh ventilation. There are two bathrooms for the end units.There is much variation in the higher floors and in the façades. The loose location and the specific orientation secure nice views of park, courtyard as well as city.

Block 8 and 9 are the two relatively low building blocks flanking the green axis square leading to the entrance of Shuangshan Park. Here we find larger apartments, 120 m² each and only two apartments per floor are being accessed from the elevator hall. The extra space gives them three bedrooms and two terraces. Here also the living room is connected with dining area which is in close relation to the closed kitchen. The kitchen and two bathrooms here also can open to the outside. Although all apartments have good orientation and sunlight access, the gables looking at the Hefei Road are particularized, and they also have an interesting view of the square and the Hefei Road.

公寓平面设计以有效利用空间为原则，同时在各个房间之间创造空间上的联系，使户型感觉更宽敞。

7、10、11 号楼为一梯四户，约 90 平方米，有通达电梯。舒适的起居室和餐厅直接相连，封闭的厨房紧邻餐厅。每户都有两个卧室和一到两个阳台，厨房和卫生间自然通风。端户为两卫。每栋楼上部平面、立面变化更为丰富，足够的楼间距和建筑体量的方向性保证了每户良好的景观性，或览山景，或看园景，或观街景。

8、9 号楼为一梯二户，高度相对较低，位于通往双山公园的绿色景观轴广场两侧。户型较大，每户面积约 120 平方米，有三个卧室和两个阳台，有通达电梯。起居室和餐厅直接相连，封闭的厨房紧邻餐厅。厨房和卫生间自然通风。尽管每户公寓都有很好的朝向和光照，沿合肥路山墙面作了特殊处理，能够观看广场和合肥路的风景。

Walls
墙面设计

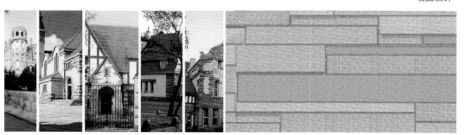

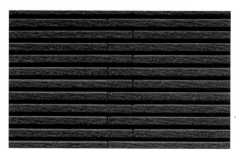

2nd Floor Plan
2 层平面图

SUSTAINABILITY 可持续性

STRUCTURE 结构

MATERIAL 材料

BALCONY 阳台

Cosmopolitan, Puerto Rico

波多黎哥大都会公寓

Architect: RTKL Associates Inc.
Client: Interlink
Location: San Juan, Puerto Rico
Gross Floor Area: 27,871 m²
Floors: 19
Photography: David Whitcomb

设计公司：RTKL Associates Inc.
客户：Interlink
地点：波多黎哥圣胡安市
建筑面积：27 871 平方米
层数：19
摄影：David Whitcomb

With RTKL's extensive portfolio in revitalization of dense, urban settings, the client Interlink Group is looking to create a new, upscale property in San Juan that would anchor and energize the once flourishing Miramar urban district. The Cosmopolitan reestablishes the standard of the Puerto Rican high-luxury market.

由于 RTKL 在密集城区改造项目上有丰富的设计经验，客户 Interlink 希望在圣胡安开发一个新的高档房产项目，从而引导并推动一度非常繁盛的米拉马尔城区的发展。大都会公寓项目重塑了波多黎哥高端奢华市场的市场标准。

The 27,871-m², 19-storey tower features 62 standard townhouse and penthouse units in the first phase. The RTKL design team nimbly embraces the local culture and tradition and enriches it with a global perspective. To achieve exclusivity and security, a resident rides an individually-keyed elevator from the garage to a private lobby at their front door. Within each residence, guest entertainment and reception is discretely separated from family living and casual dining. Modern, furniture-inspired, Poggenpohl kitchens capture a sophisticated attitude. Balcony terraces extend outdoor living for intimate family use and more formal entertainment. The laundry and day quarters are accessed by domestic staff from the service corridor and separate elevator.

To integrate all aspects of planning and building design, define the target market and clarify design objectives, RTKL provides project branding services including naming, logo design and brand application.

项目一期工程建筑面积 27 871 平方米。19 层高的塔楼包括 62 套标准联排别墅和顶层公寓。RTKL 设计组在设计中有选择地融入了地方文化和传统，并从全球视野加以深化。为了确保无他人进出和业主安全，电梯有业主所在层数限制，每个业主只能从车库到达所住层。每个单元中，客人娱乐和接待区都与住宅区以及休闲餐饮区完全隔开。厨房采用时尚的博德宝橱柜。阳台或露台可利用起来举行家庭聚会或正式的娱乐活动。内部员工从工作走廊和单独的电梯到达洗衣房和工作区。

为了整合所有规划和建筑设计、确定目标市场、明确设计目标，RTKL 还提供了项目品牌服务，包括命名、Logo 设计以及品牌应用等。

FEATURE 特点分析

STRUCTURE

To capture natural light and openness, locally requisite continuous shear wall construction is replaced with a shear wall core and perimeter columns. Hurricane shutters are succeeded by high-impact resistant glass.

结构

为了利用自然光和开放区，当地要求的连续剪力墙用剪力墙核心体和边柱代替。防风百叶采用高耐冲玻璃。

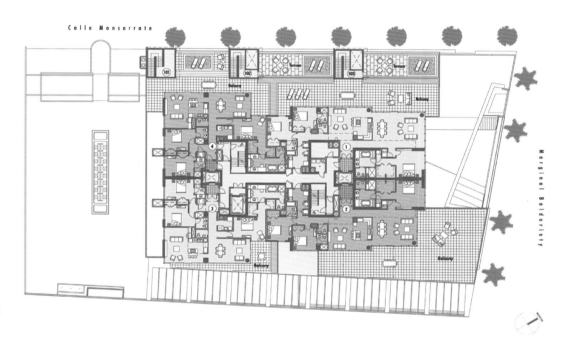

Floor Plan
楼层平面图

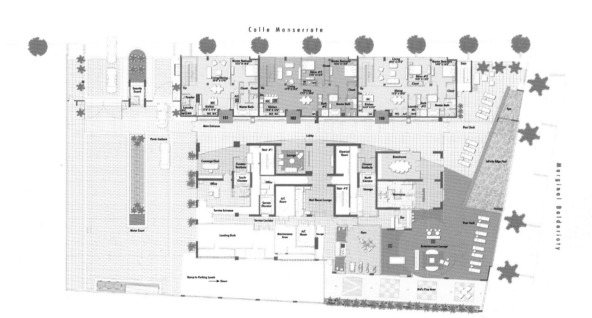

Floor Plan
楼层平面图

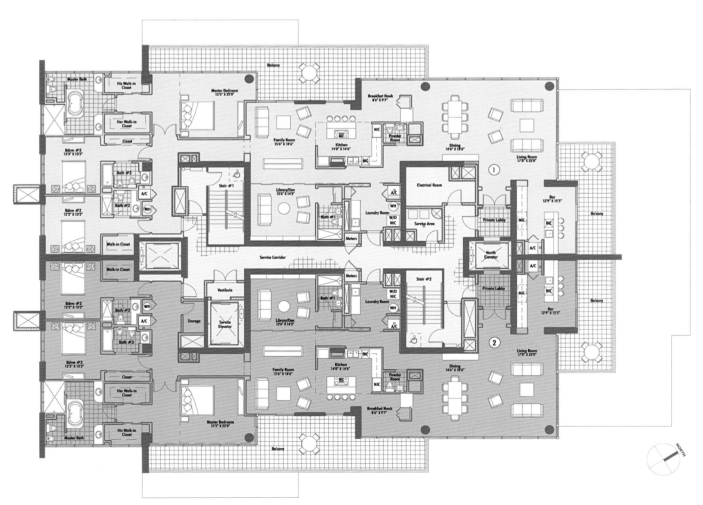

Floor Plan
楼层平面图

STRUCTURE 结构
FAÇADE 立面
MATERIAL 材料
BALCONY 阳台

Prisma, Groningen, the Netherlands

荷兰格罗宁根棱镜公寓楼

Architect: NL Architects
Client: Stichting De Huismeesters, Ed Moonen, Roelof Jong
Location: Groningen, the Netherlands
Site Area: 2,600 m²
Gross Floor Area: 8,650 m²
Floors: 16
Height: 49.5 m

设计公司：NL Architects
客户：Stichting De Huismeesters、Ed Moonen、Roelof Jong
地点：荷兰格罗宁根市
占地面积：2 600 平方米
建筑面积：8 650 平方米
层数：16
高度：49.5 米

It is surprising that in a miraculously flat country like the Netherlands only so few towers emerge. The potential of the unobstructed view—as one of the main topics in Dutch painting—is in housing under developed. There seems to be a collective fear of elevators. (Which maybe explains the success of the Dutch movie *Down*—originally *De Lift*).

在荷兰这样的平原地区出现仅有的几座高楼非常令人惊奇。毫无阻挡的广阔视野也是荷兰画作里的主要题材之一，不过这只能在高层建筑中实现。荷兰人似乎普遍恐惧电梯（这也许可以解释荷兰电影《杀人电梯》的成功原因）。

Prisma takes the view—one of the most glamorous properties of dwelling—as starting point in combination with outdoor space. Prisma is a housing project for 52 apartments in a total of 16 storeys and some additional facilities in the city of Groningen. A block of 50 m high can in the Netherlands already be considered high. Groningen is renowned for its progressive architecture. The north west area is called Vinkhuizen. It is a typical CIAM-based post-war development and is a mixture of middle-high apartments in an open layout and some semi-high rises, including a lot of greenery and a lot of public space, a lot of parking. Still the area is not particularly popular at the moment. Much of the housing is no longer considered fit for the 21st century. Prisma is part of the nationwide, large-scale renovation operation that now is underway in the Netherlands.

棱镜公寓将寓所最具魅力的特点之一的景观视为与外界空间连接的起点。棱镜公寓是格罗宁根市的一项住宅计划，它总共16层，包含52套公寓和一些配套设施。50米高的建筑在荷兰早已被认定为高楼了，格罗宁根市也因其先进的建筑而知名。西北部地区被称为 Vinkhuizen，它是国际现代建筑会议确定的战后发展的典型，也是在开放布局限定性条件下中等高度的公寓和半高层建筑的一种混和布置，其中包括许多绿色植物、公共活动空间和停车场。当前这一地区还不特别为公众所知，而许多房屋也不再被认为适合 21 世纪。棱镜公寓是荷兰全国范围内大规模建设项目之一，也是在建项目。

Locarion in Groningen
城市区位图

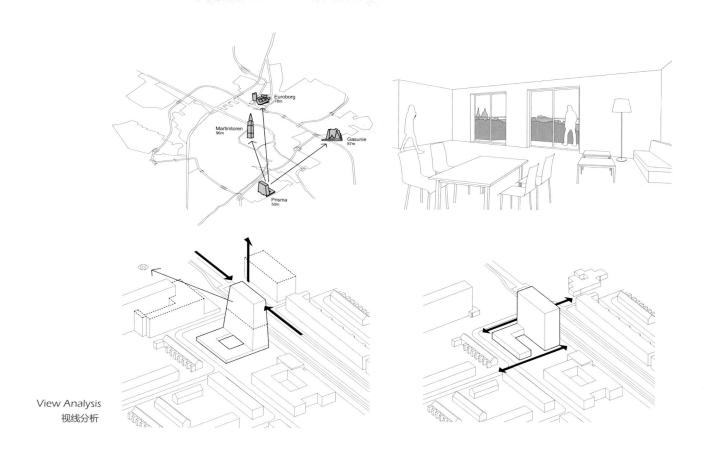

View Analysis
视线分析

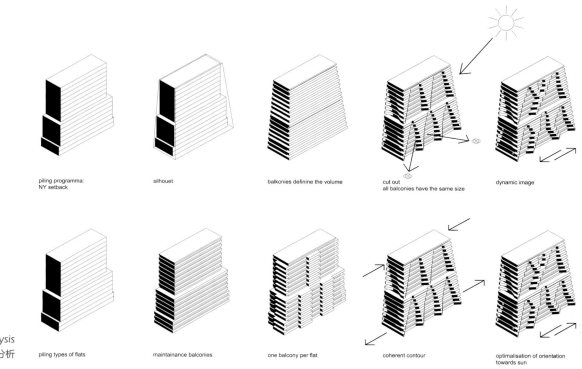

| piling programma: NY setback | silhouet | balkonies definine the volume | cut out all balconies have the same size | dynamic image |

Balcony Sunlight Analysis
阳台光照分析

| piling types of flats | maintainance balconies | one balcony per flat | coherent contour | optimalisation of orientation towards sun |

Structural Analysis
结构分析

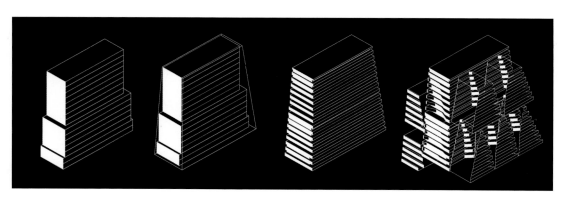

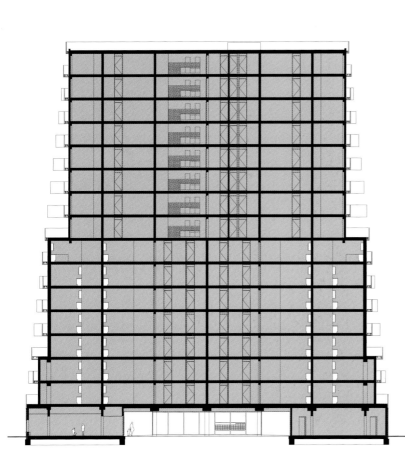

The building consists of a simple stacking of the desired apartment types. The largest ones are below, the smallest ones on top: a weirdly proportioned Maya pyramid. The contour reminds people of the archetype of the high-rise: in a way comparable to the Manhattan zoning. This rational structure serves as the backbone for a more frivolous addition. The balconies are draped around the structure almost like couture. As the first act the corners of the stepped volume are connected. The surplus of balcony that emerges as a consequence now is excavated. We take as point of departure that per category of apartments the balconies should be the same size. But the proportions change depending on the position in the block.

高楼由人们所喜欢的公寓类型集合而成,最大的户型在下面,最小的在顶部,像比例微妙的玛雅金字塔。建筑的轮廓让人联想到高楼大厦的原型,某种程度上比得上曼哈顿区。这座理性建筑物对更多零碎的附加物而言是座骨干建筑。阳台围着建筑呈打褶形布置,就像一件高级时装。作为第一步,台阶的拐角相连成一体,形成连续的造型。由此产生的阳台盈余如今被挖掘。各种户型公寓的阳台看起来尺寸都是相同的,但其比例却随所在位置有所变化。

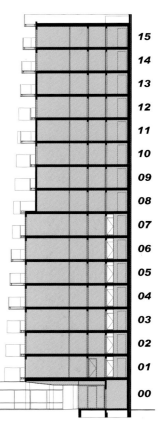

15
14
13
12
11
10
09
08
07
06
05
04
03
02
01
00

Section
剖面图

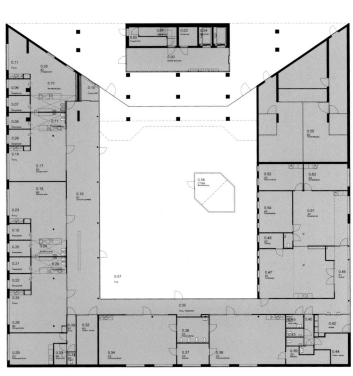

Ground Floor Plan
首层平面图

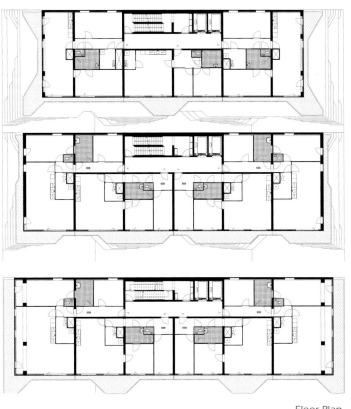

Floor Plan
楼层平面图

Facade Structure Analysis
立面结构分析图

Walkaround
外观示意图

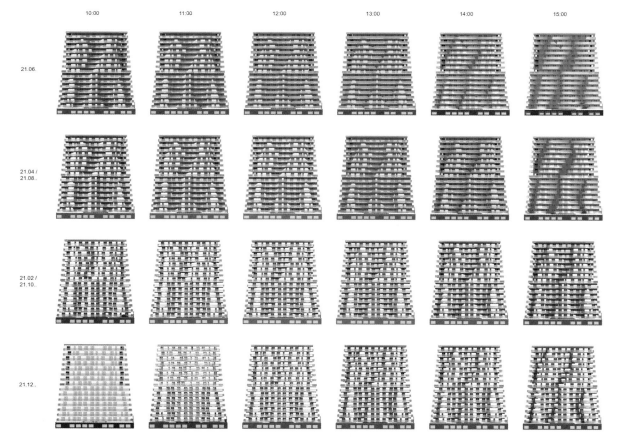

Sun Study South
南面光照研究

FAÇADE

The balconies are arranged around the building in a draped shape like a fashionable skirt. The continuous form is done by connections on conners of steps. The balconies in each unit seem to be the same size but in different proportions as the location changes.

结构

阳台围着建筑呈打褶形布置，就像一件高级时装。台阶的拐角相连成一体，形成连续的造型。各种户型公寓的阳台看起来尺寸都是相同的，但其比例却随所在位置有所变化。

The building will basically attract the elderly that hopefully will be able to spend the rest of their lives there.

At ground level, there is a nursery, children's playground and medical facilities. The entrance of these facilities and the apartments are combined in an attempt to enhance the regular dimensions that are quite limited.

这栋建筑将主要吸引老年人，他们希望在此度过剩余的生活时光。

建筑的底层设置有一个托儿所、儿童游乐场和医疗设施。这些设施的入口和公寓结合，试图提升原有的局限空间。

BALCONY 阳台
STRUCTURE 结构
MATERIAL 材料
FAÇADE 立面

De Kameleon, Amsterdam, the Netherlands
荷兰阿姆斯特丹 De Kameleon 公寓楼

Architect: NL Architects
Client: Principaal / De Key
Location: Amsterdam, the Netherlands
Site Area: 9,250 m²
Gross Floor Area: 55,500 m²
Photography: Luuk Kramer, Marcel van der Burg
Realization: 2012

设计公司：NL Architects
客户：Principaal / De Key
地点：荷兰阿姆斯特丹市
占地面积：9 250 平方米
建筑面积：55 500 平方米
摄影：Luuk Kramer、Marcel van der Burg
完成时间：2012 年

De Kameleon is a supersized housing block including a new shopping center and plenty of parking in the area formerly known as Bijlmermeer.

De Kameleon 是一座包含新购物中心和大型停车场的超大新型住宅大厦，所在的地点是以前的庇基莫米尔居住区。

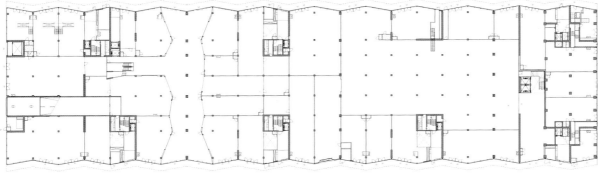

Ground Floor Plan
首层平面图

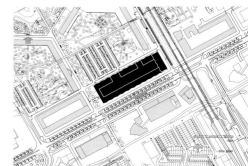

Site Plan
总平面图

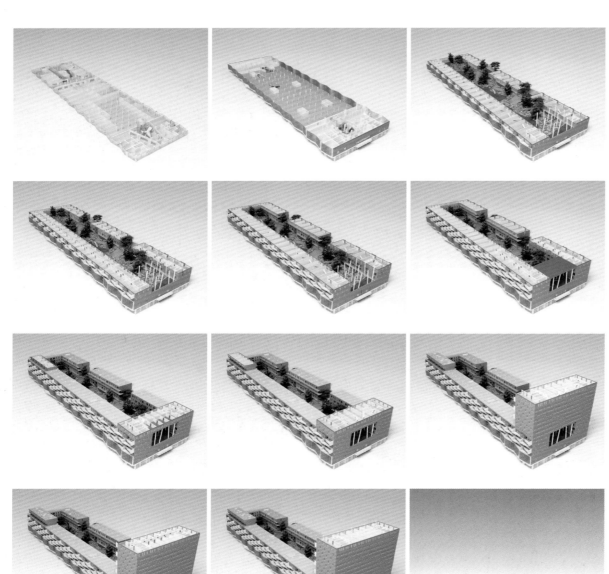

Development
建设过程

The Bijlmermeer is one area in the Netherlands that sometimes is considered as a ghetto. At the moment the area is going through a radical renovation process: an attempt is being made to turn it into a regular Dutch suburb. Standard low-rise housing is introduced that replaces the 10-storey apartment buildings but retains the green spaces in-between them. In spite of the new format, the Bijlmermeer remains exotic: it is the place to be for a sensational Roti or sundried Bats.

The Bijlmermeer features a fantastic elevated subway track, maybe the only suitable backdrop for an R&B video in the Netherlands. De Kameleon is placed along the Karspeldreef, one of the main arteries in the area. It is quite a surprise that amidst the new ideology of the small scale such a large new building is projected. De Kameleon is organized in horizontal slices. On ground floor is the new shopping center. All shops are accessible directly from public space, and there is no collective interior: De Kameleon is not a mall.

庞基莫米尔居住区是荷兰的一个地区，有时也被称为犹太人区，当前这一区域正在进行着空前的整新，正在将其改变成整齐的荷兰郊区。正在建设标准低层住宅来取代 10 层的公寓大楼，其间的绿色空间仍然被保留。尽管有新的格式的建筑，庞基莫米尔居住区还是保持了它的独特性：这里是吃烤肉和烤干蝙蝠的好地方。

高架轨道交通是庞基莫米尔居住区的特色，也许这里是荷兰唯一适合拍摄 R&B 录像的场景。De Kameleon 坐落在 Karspeldreef 旁边，这里是这一地区的主要交通通道之一。在新兴的小规划观念和大型新建筑的建造中保持平衡，这相当令人惊奇。De Kameleon 按水平分层布局，底层是新建的购物中心，从公共空间能直接到达所有的商店，没有共同的内部空间：De Kameleon 不是一个购物商场。

A 10-storey slab with 58 apartments rests on this flat block. It creates a counterpoint to the horizontality and becomes a "billboard" facing the subway. A supersized window visually connects the elevated subway and the elevated garden that are precisely the same height.

The supermarket, normally a bulky program with extensive impenetrable façades is embedded in smaller units that as such both differentiate and activate the "plinth". There is one shortcut, the passage, at two-thirds of the length, creating an "8". The "8" is good for circulation and good for business. From here an escalator connects to the next level, continuing the 8 in the 3rd dimension. On the 2nd floor is one more supermarket, easily accessible from the public parking on the same level.

公寓高 10 层，共有 58 套住宅，坐落在这平坦的街区，它为街区创造了水平状态的对位，成为面向地铁的巨大广告牌。一个巨大的窗户形象地连接起高架铁路和空中花园，两边的高度刚好相同。

超市体量很大，外立面宽大坚固，内部被分为小型单元，以明确分区并为底层带来活力。其中有一条捷径——走廊，在三分之二长度，处形成 "8" 字形。"8" 字型有利于交通循环，有利于商务交流。这里有一部自动扶梯连通到下一层楼，在第三维度上延续了 "8" 字形。第二层是另一个超市，从同一层的公共停车场就能直接进入。

East Façade
东立面

South Façade
南立面

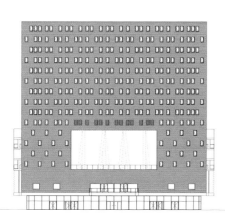

West Façade
西立面

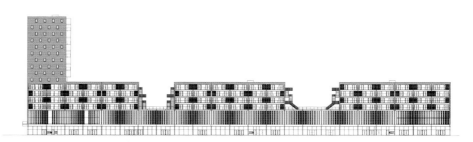

North Façade
北立面

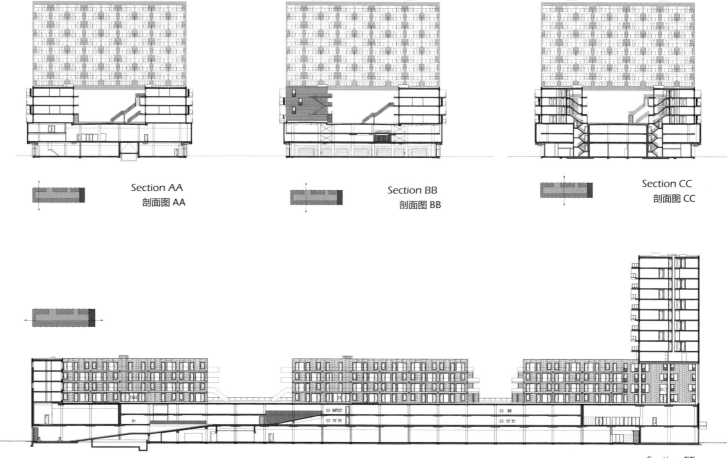

Section AA
剖面图 AA

Section BB
剖面图 BB

Section CC
剖面图 CC

Section FF
剖面图 FF

Housing Type D3
户型图 D3

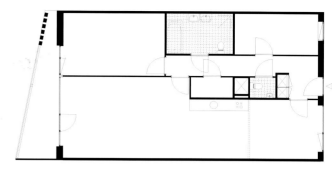

Housing Type C1
户型图 C1

8th, 10th, 12th, 14th, 16th Floor Plan
8、10、12、14、16 层平面图

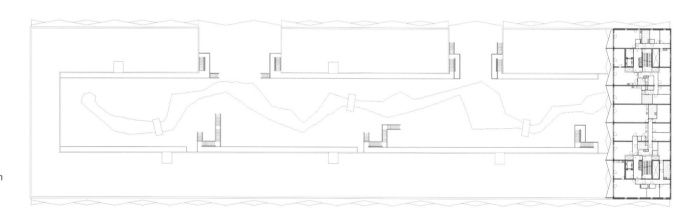

9th, 11th, 13th, 15th Floor Plan
9、11、13、15 层平面图

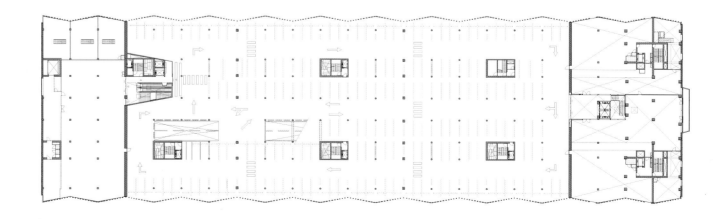

1st Floor Plan
1 层平面图

FEATURE 特点分析

STRUCTURE

The repetitive structure makes the project affordable. The rhythmic building bays of 8 m and the parking and shopping grids correspond nicely. Every other carrying wall is extended to support the balconies and to provide privacy. The large balconies create dynamic patterns. Winding stairs lead to the garden and differentiate the large courtyard.

结构

复式结构让工程项目更有效，规律的8米建筑隔区、停车场、商店布置井井有条。其它承重墙都延伸出来，以支承露台，保护隐私。大型露台形成了动态分布，盘旋楼梯直通花园，使大庭院区分开来。

Positioning the parking on top of the shops is proofed to be cheaper than in a basement. The parking is "charged" by the supermarket on one end and food court and fitness center on the other. Since these programs feature large floor to ceiling heights, an extra parking level fits in. The residents will park their cars here.

The façade of the parking is open to the sides allowing natural ventilation. A very large garden is placed on top of the parking. It includes 12 series of trees and a river! The garden is surrounded by a 4-storey housing block containing 168 apartments. The side facing the Karspeldreef is continuous to protect the garden from street noise and to create an "urban wall"; the other side facing the typical hexagonal green space is punctured. The gaps can be used as playgrounds and for BBQ.

在商场的顶部设立停车场，成本比在地下室要低。停车场的一端是超市，另一端是美食街和健身中心。由于这些项目占用了大量地面和室内净空，有一个附加的停车层配套在里面，住户的车就停在这一层。

停车场的正面敞开，以利于自然通风。停车场顶上设置了一个很大的花园，种植了 12 个种类的树木，有一条小河。公园四周环绕着一座四层的住宅楼，它有 168 套公寓。公寓朝向 Karspeldreef 的侧面一直平行延续，作为一座"城市墙"以保护花园免受街道噪音污染；另一边朝向六角形绿色空间的方向被打通，开口处的空地用于居民游乐和烧烤。

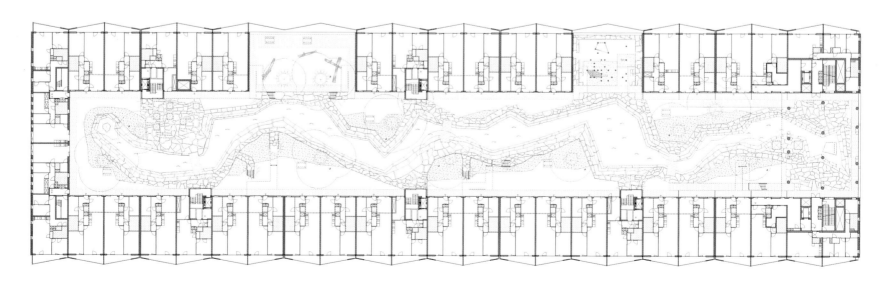

3rd Floor Plan
3层平面图

VIEW 视野

STRUCTURE 结构

SHAPE 造型

FAÇADE 立面

Black Prince Road, London, UK

英国伦敦黑太子路大厦

Architect: Keith Williams Architects
Client: Ristoria Ltd
Location: London, UK
Site Area: 11,160 m²
Realization: 2012

设计公司：基思·威廉姆斯建筑事务所
客户：Ristoria Ltd
地点：英国伦敦市
占地面积：11 160 平方米
完成时间：2012 年

Black Prince Road is the firm's first tower project and is located close to the River Thames in London's Vauxhall. The highly complex project is sandwiched between the railway viaduct into Waterloo and the tight-knit Victorian street pattern that surrounds it.

黑太子路大厦是该公司研发的第一个建筑项目，靠近英国沃克斯豪尔泰晤士河。这样高度复杂的项目夹在通入滑铁卢的铁路高架桥和围绕着它紧密的维多利亚街之间。

The development site is currently occupied by a redundant 7-storey 1950s office building on an irregular plot. The tower will contain a total of 101 private and social sector flats, with commercial development within the base storeys. The project is affected by protected strategic views and has been carefully modelled to ensure it will make an elegant contribution to the London skyline. The building, although set back one block from the river, will offer spectacular views of the Houses of Parliament from the 7th floor and above.

The building's surfaces are solid with punctuations for the syncopated pattern of the window openings giving a rich façade texture. At the uppermost levels, the building becomes more heavily glazed with a crystalline element forming the junction of the building with the sky.

On the plan, the tower's façades are split into two embracing plates, which encapsulate the glazed body of the building. The meeting point of the two plates occurs at the clipped-off corners of the parallelogram plan where a series of cascading balconies form a "zipper" which runs up each leading edge of the tower. At the cranked corners of the plates, stepped balconies appear as "knuckles" to articulate these key corners.

The floor plates in the tower repeat in groups of three and generally the window locations are the same on each of the three levels in a set, before shifting on the set above. This generates strips of tall, slender glazing panels, which are dispersed in a semi-random pattern across the solid plate surfaces as an applied design. The windows are proud of the plane of the wall to aid construction and to modulate the wall surface.

The completed tower will offer spectacular views out to the River Thames, the Houses of Parliament, the City, Canary Wharf and the South Downs beyond.

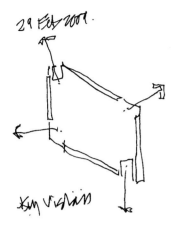
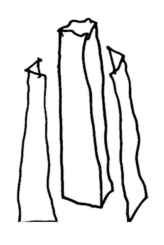
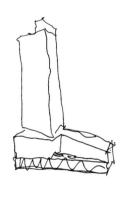
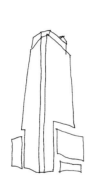
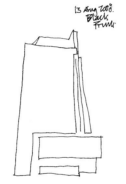

Sketch
手绘草图

目前该地区被一栋多余的建于一块不规则地块上的 19 世纪 50 年代的 7 层办公楼所占据。该建筑将包含 101 间私人和公共公寓，底层为商业区。该项目受到规划景观的影响，必须精心地塑造以确保它成为伦敦天际线的一个优雅景观。尽管建筑与泰晤士河相隔一条街区，但是站在建筑的 7 楼以上就可以饱览议会两院美丽的风景。

建筑表面的立体感和开口为间隔的窗口结构提供了丰富的立面纹理。在建筑的最高层，建筑融入水晶元素，使其釉面感加重，以形成连接天空与建筑之间的枢纽。

在规划中，建筑的立面被分成两个相互拥抱的楼板，并将大楼的釉杯封装在内。两个楼版的汇聚点出现在平行四边形的边角处，那里由一系列层叠的阳台形成 "拉链" 状，沿着建筑的每个前沿部展开。在楼板的拐角处，阶梯式的阳台如同关节一般成为建筑的关键元素。

建筑的楼面板每三个为一组，重复出现，同时窗口的位置在打开之前也大致相同。这样通常形成身材高挑的玻璃面板，并作为一种应用设计分散在楼板表面的半随机模式中。窗户是墙面设计来支持墙体建设和调节墙面的成功之处。

完成后的建筑将饱览泰晤士河、议会两院、城市、伦敦金融区和南部唐斯的亮丽风景。

FEATURE 特点分析

STRUCTURE

The parallelogram plan combined with a vertical taper gives the building a blade-like appearance. The taper helps maximise daylight into the existing residential accommodation along Salamanca Place, and makes the tower more slender by introducing an exaggerated perspective from whichever side the building is viewed. Sitting between the towers at Westminster House and the recently consented Hampton House, the building will add verticality to the silhouette of the river frontage that currently appears monolithic.

结构

项目的平行四边形规划配上垂直锥形的结构让建筑外观整体看起来像一个刀片。这样的锥形构造在白天可以让沿着萨拉曼卡广场的住宅楼最大限度地吸收阳光，同时又使得建筑不管从哪个角度看都很苗条。坐落在威斯敏斯特住宅和最近新建的汉普顿住宅之间的建筑，将为该建筑面向河岸线一侧添加一种垂直的轮廓感。

Model
模型图

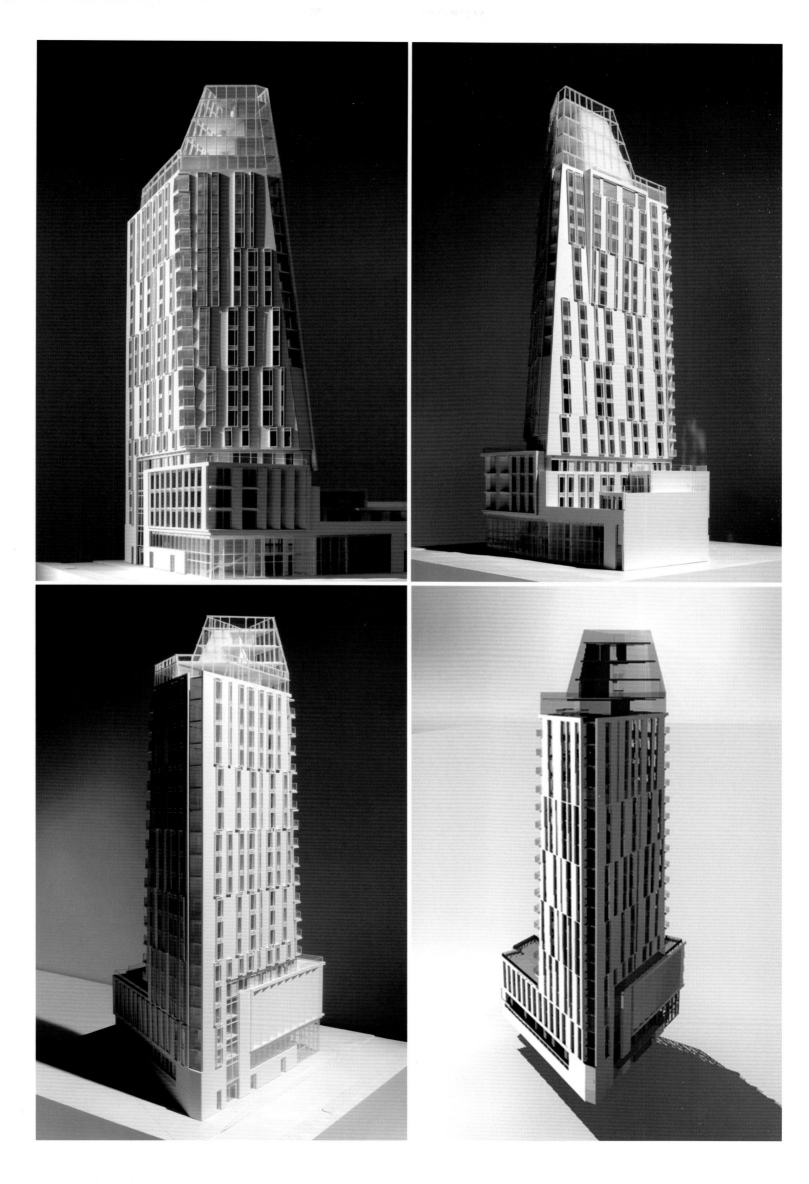

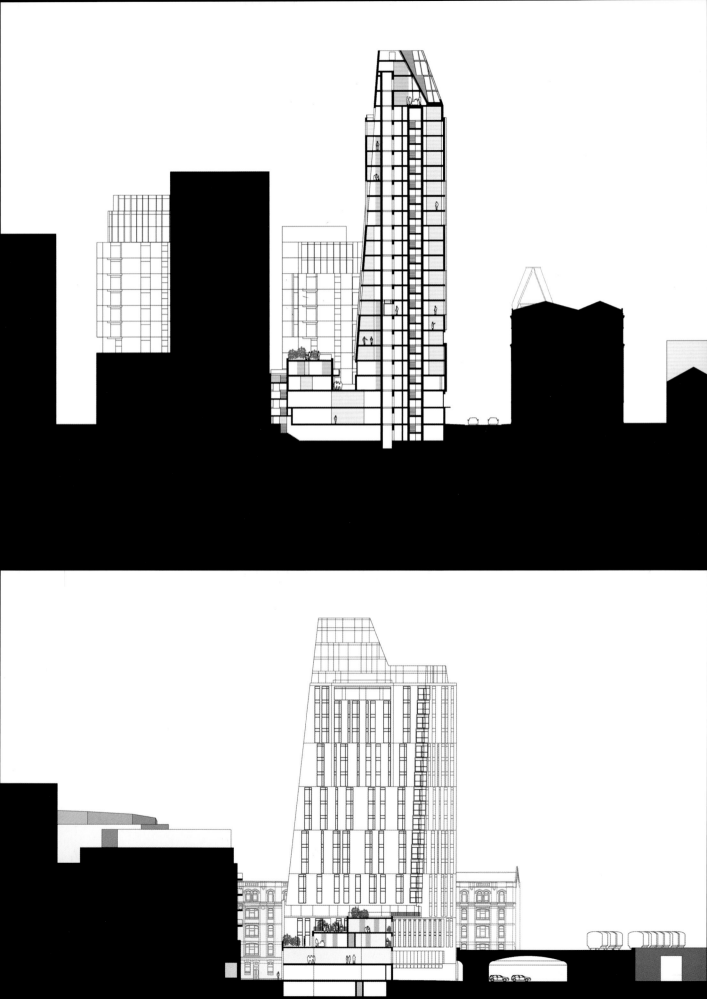

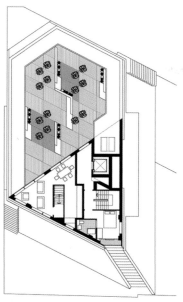

22nd Floor Plan
22 层平面图

20th Floor Plan
20 层平面图

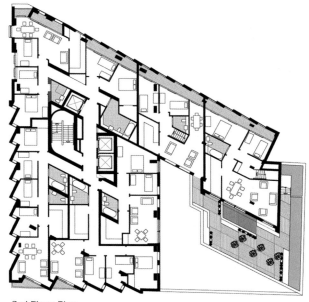

3rd Floor Plan
3 层平面图

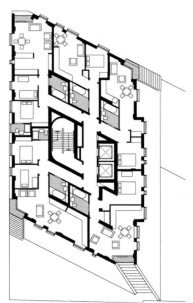

17th Floor Plan
17 层平面图

2nd Floor Plan
2 层平面图

WINDOW 窗口
STRUCTURE 结构
SHAPE 造型
FAÇADE 立面

Tirana Station Towers, Albania
阿尔巴尼亚地拉那车站公寓

Architect: de Architekten Cie.
Client: Edi Rama, Mayor of Tirana
Location: Tirana, Albania
Gross Floor Area: 42, 000 m²
Floors: 23
Realization: 2012

设计公司： de Architekten Cie.
客户：地拉那市长埃迪·拉马
地点：阿尔巴尼亚地拉那市
建筑面积：42 000 平方米
层数：23
完成时间：2012 年

The design for these two towers is a blend of imagination and realism. The towers present a strong and recognizable image without lapsing into the inflation-prone pursuit of "iconic" expression.

项目的设计融合了想象和现实主义的格调。在不倒退到盲目地追求"经典"表达的前提下实现了一个令人印象深刻又独具特色的建筑。

The project demonstrates the combination of classic and modern perfectly. The shapes of two buildings feature uniquely but similarly, integrated with surrounding landscape harmoniously. Ornate façade not only gives a luxurious sense to the buildings, but also clearly outlines interior spaces. It is a truly high combination of indoor and outdoor spaces in the same building.

The outward form is distinctive yet restrained enough not to exclude or overwhelm other architecture in the immediate surroundings. Construction and finishing are based on locally available technology, creativity and craftsmanship, with an emphasis on the use and encouragement of ecological technology.

Both towers contain a maximum mix of offices, hotels and apartments, supplemented by large, communal voids in tower A and public facilities in tower B. At ground floor level, the similarly public retail programme continues into a below-grade level.

In tower B, this lower level links up with the neighbouring park and contains bars, restaurants and a music pavilion. The public programme in the two towers underscores the ambition to offer this key city location something more than the generally private, indoor world of living and working, thus invigorating this part of Tirana.

In one of the towers the programme is horizontally mixed, in the other vertically. In both cases, the room for manoeuvre allowed by the prescribed envelope is used to stagger the façades and so gives each tower a distinctive vitality. Thanks to their height, they will also offer a variety of superb (horizontally or vertically framed) views of the city, the new park and the surrounding mountains.

Another thing the towers have in common is the structural setup and the flexible floor plans, allowing changes to the programme, now or in the future. This underlines the essence of this design proposal: an architecture that is simultaneously specific and generic.

项目建筑完美地展现了经典与现代的结合。两栋建筑造型独特，却有共性，同时又与周围景观和谐共融。华丽的立面不仅渲染了建筑的奢华，也将室内空间清晰地描绘出来，真正做到了建筑室内外空间的高度融合。

建筑外观独特，但严格的限制并没有使其排除或是淹没周围的其他建筑。建设和修整基于本地现有技术、创造力和工艺，而重点是对生态技术的提倡和使用。

建筑最大限度地混合了办公、酒店和公寓功能，并由 A 楼提供的大型公共空间和 B 楼的公共设施以作补充。在建筑的首层，类似的公共零售项目设施一直延伸到地面下层。

在 B 楼中，底层部分包括酒吧、餐馆和音乐厅，并同附近公园紧密相连。两栋建筑的中公共区域为这个关键的城市位置提供了更为隐私的室内生活和工作空间，让这个区域更为繁荣。

其中一座建筑呈水平向混合发展，而另一座则呈垂直向发展。两种情况下，这两栋体量都采用一种特制的墙面，以方便允许对房间进行调制布局。这种墙面使建筑立面体现出一种交错摇晃的感觉，并赋予它与众不同的活力。由于建筑的高度，可将来自城市、公园和周围山脉的（水平或垂直框架的）景观尽收眼底。

建筑的另一个共同点是建筑的结构和灵活的地面设计，它允许在某个时间被更改以方便建筑现阶段或将来做调整。这突显了方案设计的本质：兼具特殊与一般的建筑构造。

FEATURE 特点分析

STRUCTURE

The two towers are different but unmistakably related; like brother and sister they share the same DNA. The façades of both towers are composed of the same façade elements, with a functionally determined ratio of open to closed which makes the internal spatial organization legible on the outside.

结构

两栋建筑之间存在着差异，却又有明显的联系，就像哥哥和妹妹共享相同的 DNA 一样。两栋建筑的外立面由相同的元素组成，并通过一定的开关比率让室内空间组织可以在外部被清晰地辨认出来。

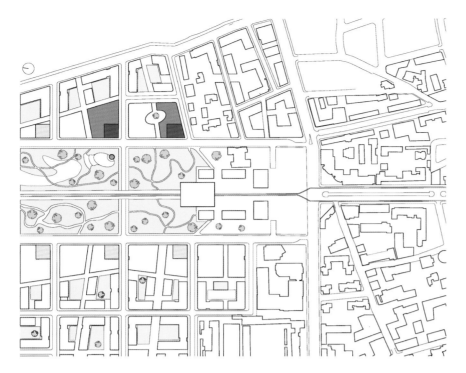

Site Plan
总平面图

Location Plan
区位示意图

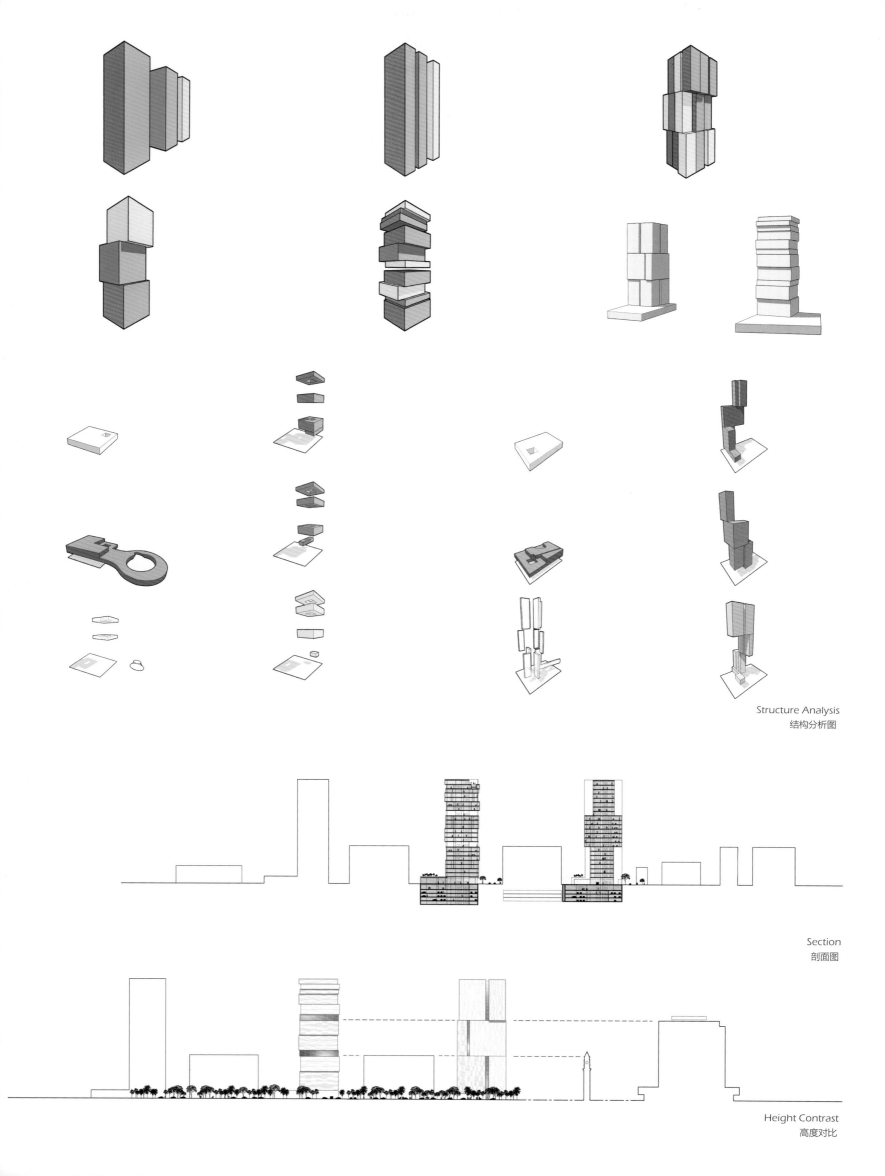

Structure Analysis
结构分析图

Section
剖面图

Height Contrast
高度对比

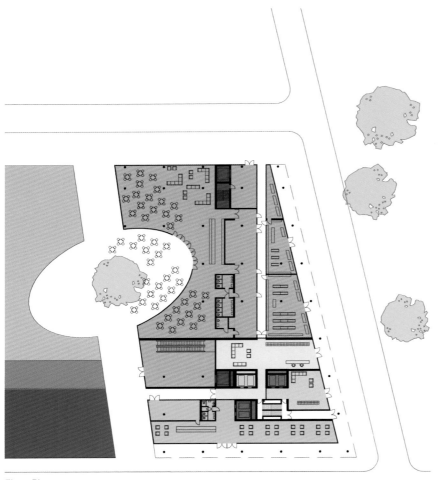

Floor Plan
楼层平面图

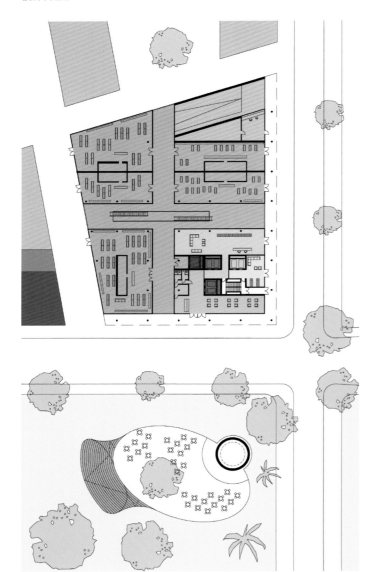

Floor Plan
楼层平面图

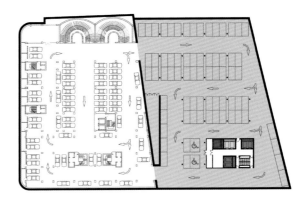

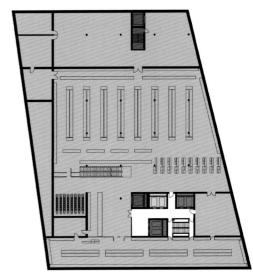

Floor Plan
楼层平面图

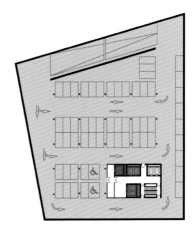

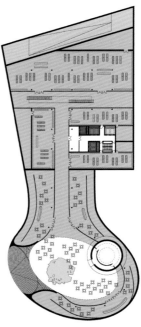

Floor Plan
楼层平面图

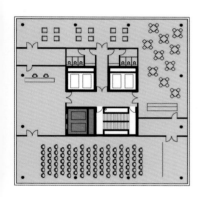
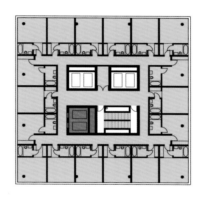
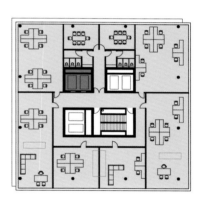

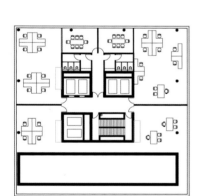

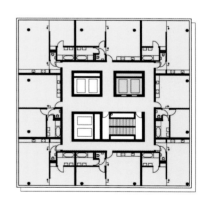

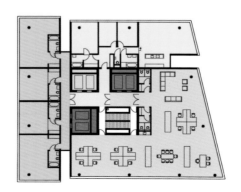

Floor Plan
楼层平面图

SHAPE 造型

STRUCTURE 结构

TERRACE 露台

FAÇADE 立面

North-Milsons Point, Sydney, Australia

澳大利亚悉尼北米尔森斯公寓

Architect: PTW Architects
Client: Australand / Rebel Property Group
Location: Milsons Point, NSW, Australia
Site Area: 1,074 m²
Floors: 15
Height: 57.66 m
Photography: PTW Architects, Brian Steele

设计公司：PTW Architects
客户：Australand / Rebel Property Group
地点：澳大利亚新南威尔士州米尔森斯区
占地面积：1 074 平方米
层数：15
高度：57.66 米
摄影：PTW Architects、Brian Steele

North Residences is a conversion of an existing 1960's commercial building (Eagle House) into a luxury apartment building adjacent to Sydney Harbour.

本案公寓是由建设于 20 世纪 60 年代的商业建筑转化而来，是一栋毗邻悉尼海港的奢华公寓大楼。

The building takes full advantage of the location by maximising views across the harbour, city skyline and surrounding environment. The design of the building derives inspiration from Sydney's port environment—the air, wind and water—and its diverse international population. Viewed from across the bay, the glass-clad towers rise from a sculpted base to resemble a huge sculpture. At 15 storeys tall, the building prominently carves out the leading edge of Sydney's emerging new skyline.

The development contains 76 spacious apartments planned in a variety of configurations over 15 levels, ranging from studios to one-bedroom, two-bedroom and three-bedroom apartments. Basement parking is provided for 66 cars. Commercial and retail space occupy the ground floor.

Designed to advance Sydney's leadership as a global "Live Work Play" destination, the building—the first Marina Bay project to introduce residential units—contains restaurants, health clubs, recreation decks with pools and car-park facilities. To create a true 24/7 lifestyle environment in the port of Sydney, and to help reduce traffic congestion and smog, the building is designed to directly access the MRT and park amenities along the waterfront. Retail, dining and work environment are all within an easy walk.

The distinctive form of the building and prominent location have established a memorable landmark on the north shore of Sydney Harbour.

FEATURE 特点分析

STRUCTURE

Existing reinforced concrete structure with reinforced concrete slabs alterations and additions with paint finish and aluminium channel edge trim. Infill aluminium glazing with curved glass projecting bay windows. Steel framed aluminium panel clad awnings and operable pergolas. Louvre wall cladding to plant enclosure.

结构

现有的钢筋混凝土结构、改建及加建的混凝土板、漆涂表面、铝制修整边缘是建筑的主要结构，填实的铝制釉面和以曲面玻璃为材料的凸窗格外引人注目，还有钢架铝制面板覆层的遮阳棚和可拆卸的藤架。并以植被从外部将其围绕起来。

建筑充分利用位置优势，将港口、城市天际线和周围环境收进囊中。同时，这一设计基于悉尼港口怡人的环境——蓝天、白云、清新的空气、水和多元化的人口元素。从滨海对面放眼望去，巨型玻璃塔从一个雕刻底座上拔地而起，恰似一座巨大的雕塑。项目建凭借 15 层楼的高度，为港口勾勒出新的轮廓线。

整栋大楼 15 层的楼体包含了 76 间宽敞的公寓，从工作室到一卧室、两卧室和三卧室公寓。地下停车场可容纳 66 个停车位。地面层由商业和零售空间所占据。

这一设计旨在将悉尼打造成为一个国际领先的"生活、工作、娱乐"目的地，成为首个推出住宅单元概念的海滨湾建筑项目，包括餐厅、健身房、游泳池娱乐区、停车场等设施。为了在悉尼港口真正实现全天候不间断的生活服务模式，并为了减少交通流量、交通阻塞和污染烟雾，住宅建筑项目将交通运输系统和停车场安排在了水边。商店、餐饮和办公设施也都在附近，为居民提供极大便利。

这个独特的建筑模式和突出的地理位置俨然成为悉尼海港北部海岸的里程碑式标志。

SHAPE 造型

STRUCTURE 结构

ROOFTOP GARDEN 屋顶花园

VIEW 视野

Alba, Singapore

新加坡阿尔芭公寓

Architect: Ong&Ong Pte Ltd

Client: Far East Organization Centre Pte Ltd

Location: Singapore

Gross Floor Area: 3,852 m²

Realization: 2015

设计公司：Ong&Ong Pte Ltd

客户：远东机构

地点：新加坡

建筑面积：3 852 平方米

完成时间：2015 年

Alba is a freehold development located at 8 Cairnhill Rise, in District 09, Singapore, minutes away from Orchard MRT Station and Newton MRT Station. The expected completion date is in 2015 and it will comprise 50 units. Alba is close to the famous Newton Hawker Centre and Fort Canning Park. It is also within walking distance to the Orchard Road shopping belt.

本案是一个拥有永久地契的公寓住宅项目，位于新加坡市区第 9 区的经禧坡 08 号，距离乌节路捷运站和纽顿捷运站只需几分钟路程。项目预计将在 2015 年竣工，届时将拥有 50 套住宅单元。项目靠近著名的纽顿美食中心和福康宁公园，步行即可去到乌节路购物带。

The project is located in the prime district of Singapore urban area, and with this excellent location, it becomes the nearest project in the Orchard Road shopping district from the city CBD. It's only five minutes away from here to the Orchard Road shopping belt. It is designed by architectural design company Ong&Ong Pte Ltd, which has always been internationally award-winning, and is known as "the myth of world architecture". Alba will be the new landmark of the city. The apartment is close to Orchard station, near the Newton station and Somerset station, and it can be said that the traffic is very convenient. Anglo-Chinese Junior College (Newton), Chatsworth international school, National Junior College project also surrounding, the educational resources are excellent.

该项目位于新加坡市区黄金地段，有着绝佳的地理位置，是乌节路商圈中离城市 CBD 最近的项目，只需步行 5 分钟便可到达繁华的乌节路购物带。阿尔芭公寓由在国际上屡获殊荣、被誉为"世界建筑神话"的 Ong&Ong Pte Ltd 设计，将成为城市的新地标。公寓紧邻乌节地铁站、靠近纽顿地铁站和索美赛地铁站，可以说出行交通非常便捷。英华初级学院（纽顿）、直茨沃斯国际学校、国家初级学院也都在项目周边，教育资源非常优越。

FEATURE 特点分析

STRUCTURE

The unique Y-shaped design provides multiple options for home buyers. In addition, the characteristic apartments also raise the luxurious sense to a new standard. With outdoor sofas, pavilions, waterparks, people can enjoy the beautiful scenery at their spare time and fully immerse in the life like in a fairy world.

结构

建筑独特的 Y 型设计为购房需求提供了多重选择，特色公寓又将这种奢华生活上升到了一个新的高度。户外的沙发、凉亭、水景公园让人们在闲暇的时候还可以欣赏到美丽的风景，完全沉浸在仙侣般的生活里。

Alba has 50 freehold luxury apartments, and there are three kinds of rooms in it, 170 m² with three bedrooms, 190 m² with three plus one rooms and 210 m² with four rooms. Based on their unique construction method, three characteristic apartments will be provided on each storey. The distinctive Y-shaped layout fusing creative design of modern luxury and space gives households more living space and independence.

The design made an outdoor extension for the project, while placing the sofa and coffee table for people to rest to relax. In addition, the arched structure was shrouded over it, as the sunny shade, the rainy cover. Waterscape garden, leisure pool, funny pavilion, barbecue platform, leisure platform, gym, etc are all available, which confirms the message, "Alba, always discovering more."

阿尔芭公寓拥有 50 套永久地契豪华公寓，现有的房型为三房式 170 平方米左右，三加一房式 190 平方米左右，四房式 210 平方米左右，并以其独特的建筑方式在每一层提供了三套特色公寓。独具匠心的 Y 型布局，融合了现代式的奢华和空间运用的创新设计，给住户更广阔的生活空间和独立性。

项目设计了户外延伸部分，同时放置了沙发和茶几供人们休息放松。另外，拱形构造笼罩上空，晴天遮阳，阴天遮雨。水景花园、休闲泳池、休闲凉亭、烧烤亭、休闲平台、健身房等一应俱全，正印证了它的一句宣传语，"在阿尔芭公寓，总能不断发现更多。"

SHAPE 造型

SUSTAINABILITY 可持续性

MATERIAL 材料

ENERGY SAVING 节能

Rihan Heights, Abu Dhabi, UAE
阿联酋阿布扎比日瀚顶点

Architect: SPARK
Client: Capitala (Mubadala CapitaLand Real Estate Company LLC)
Location: Abu Dhabi, UAE
Gross Floor Area: 152,576 m²
Photography: Lin Ho

设计公司：SPARK
客户：Capitala（穆巴达拉嘉德置地房地产有限责任公司）
地点：阿联酋阿布扎比
建筑面积：152 576 平方米
摄影：Lin Ho

Rihan Heights is the first of a series of developments in the prestigious Arzanah masterplan situated at the gateway point to Abu Dhabi Island. The developments' high-end design character responds through its form, careful choice of materials and disposition to the unique location, the climate and its residential use.

日瀚顶点是著名的阿尔达奈岛总体规划开发系列中的首个项目，位于阿布扎比岛入口处。其外观、材料的精心选择、地理位置、气候以及居住功能的独特安排尽显高端设计品质。

Five residential towers seemingly float above lush podium gardens, enclosing and embracing the landscape, creating a private environment for the residents. The 5 buildings combined with 14 generous villas provide a variety of apartment types from single-bedroom apartments to penthouse units, all of which are offering spectacular views across the gardens, the city and seafront.

The design target of Rihan Heights is to build an exciting contrast between the arid desert and lush vegetation. To this end, the landscape design takes into account the natural environment of Abu Dhabi, and reduces the demand of water for irrigation. As the first phase of Arzanah master plan, here coveres a variety of landscape, particularly in the arid desert land. Vertical sky gardens form an inseparable part of every residential building, and form an eye-catching design on lacquerware, which is very interesting from height.

5座住宅楼仿佛漂浮于茂盛的平台花园之上，坐拥绝佳景色，为居住者营造出私密的环境。5座住宅楼以及14栋别墅提供了各种房型，从独间卧室公寓到屋顶公寓一应俱全。不管身居何处，都可领略花园、城市和海边的迷人风景。

日瀚顶点的目标是在干旱的沙漠和葱郁的植被之间形成一个激动人心的对比。为此，景观设计考虑到了阿布扎比的天然环境，并减少了灌溉用水的需求。作为阿尔达奈岛总体规划的一期工程，这里覆盖了多种景观，尤其是干旱的沙漠地块。垂直的空中花园形成了每一座住宅楼不可分割的一部分，并组成了醒目的图案花纹，从高处看十分有趣。

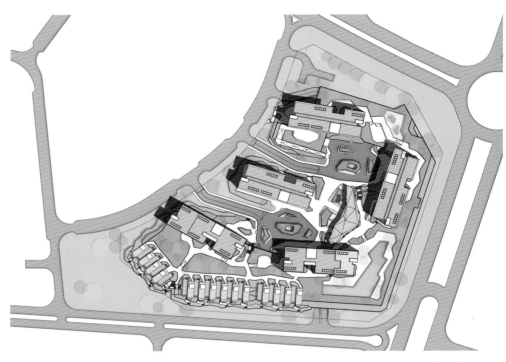

Site Plan
总平面图

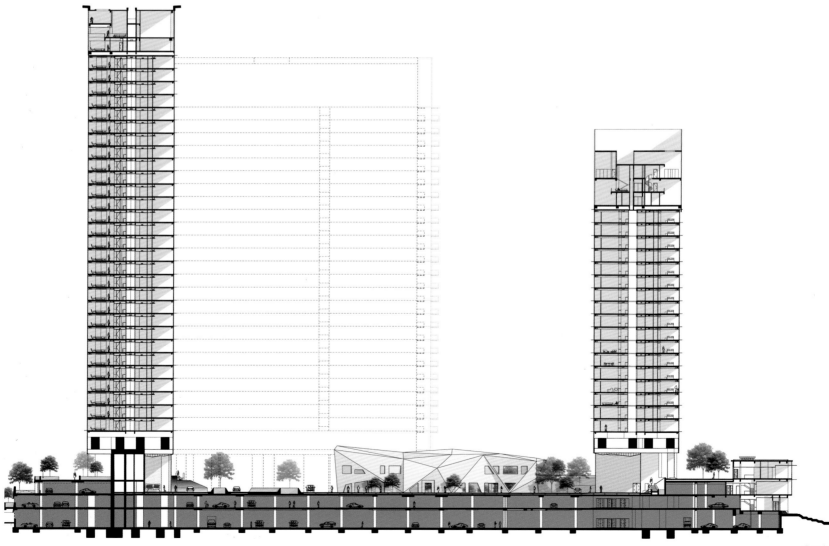

Section
剖面图

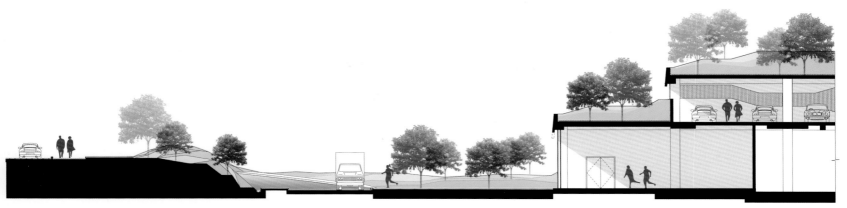

Squat Block Section
裙楼剖面图

FEATURE 特点分析

SUSTAINABILITY

Architectural design takes into account the surrounding environment fully. Careful selection of materials and unique appearance add gorgeousness and softness for the region with dry condition. Drought-tolerant vegetation conforms to local ecological condition, and creates a perfect combination of architecture and ecology, which makes the building look both high-end and environmental friendly.

可持续性

建筑设计充分考虑到周边环境，精心的选材、独有的外观为该地区干燥的环境增添了一份绚丽与柔美。耐旱的植被符合当地的生态条件，完美地打造出建筑与生态的结合，让建筑看起来既高端又环保。

Protection and utilization of water resources is also a driving force in the design, and the nature shelter made up by tree cover and other vegetation will improve the micro-climate of the massif, forming a livable and walking environment. To that end, here plants the Eastern Mediterranean and subtropical plants, close to the region's climatic conditions and requiring less irrigation.

Three-storey squat block is surrounded by unique landscape, and it contains some features such as car park. Landscape on the slope constitutes a visual pleasure, and residential building is rooted in the landscape. The buildings close to the club and other community buildings provides a kind of private city garden for residents. Housing types range from one bedroom to three bedrooms, including villas and so on. All the residential buildings will have vision of gardens or the beach and urban scenery.

水资源的保护和利用也是设计中的推动力，大树和其他植被构成的天然遮挡将改善地块的微观气候，形成了宜居和可步行的环境。为此，这里种植了地中海和亚热带的植物，接近该地区的气候条件，并且无需过多灌溉。

三层楼的敦座用独特的景观围合，容纳了停车场等功能。斜坡上的景观构成了视觉上的愉悦感，而住宅楼就扎根在这些景观中。靠近俱乐部和其他社区的建筑为居民提供了一个私人城市花园。住宅的户型从一间卧室到三间卧室不等，包括别墅等其他类型。所有的住宅都拥有能看到花园或是海滨和城市的广阔视野。

Floor Plan
楼层图

Indoor hall is solemn, the wall with simple gray tone is single but fashionable, and towering space seems clear but not empty. The lighting hanged in the middle of it, grace and effulge, adds an hot enthusiasm for this calm space. The staircase with old wood color corresponds to the geometric shape of the white wall, making the whole space look solid and art, letting classical and modern have the perfect combination.

室内大厅庄严肃穆，简单的灰色调墙壁单一却不乏品味，高耸的空间敞亮却不空旷，中空悬吊的灯饰高贵优雅、闪闪发亮，为这一份冷静的空间增添了一份如火的热情。古木色的楼梯阶配上几何形状的白色墙壁，让整个空间看起来既立体又艺术，将古典与现代做了最完美的结合。

SHAPE 造型

SUSTAINABILITY 可持续性

STRUCTURE 结构

WINDOW 窗口

Golden Gate, Amsterdam, the Netherlands

荷兰阿姆斯特丹金门大厦

Architect: arons en gelauff architecten
Client: ontwikkelingscombinatie Pontsteiger
Location: Amsterdam, the Netherlands
Height: 238 m

设计公司：arons en gelauff architecten
客户：ontwikkelingscombinatie Pontsteiger
地点：荷兰阿姆斯特丹市
高度：238 米

The location near the timber docks Houthaven is a hub within the water city of Amsterdam, and the ferry is to become an important link between the new municipal districts. The building on the Pontsteiger expresses a new élan in the relationship between these neighbourhoods. A contemporary city gate at this junction in the IJ is a fitting form for this.

项目位于水城阿姆斯特丹中心，临近 Houthaven 码头，轮渡成为市政辖区之间联系的重要纽带。位于 Pontsteiger 区的本案建筑为临近建筑群之间传达了一种活力。一座现代的城市大门屹立于 IJ 中的枢纽点，并以其恰到好处的造型设计成为该区最相称的建筑。

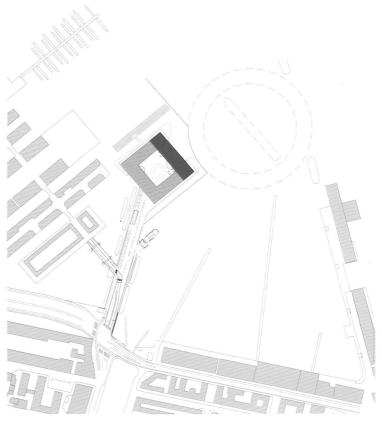

Location Plan
区位平面图

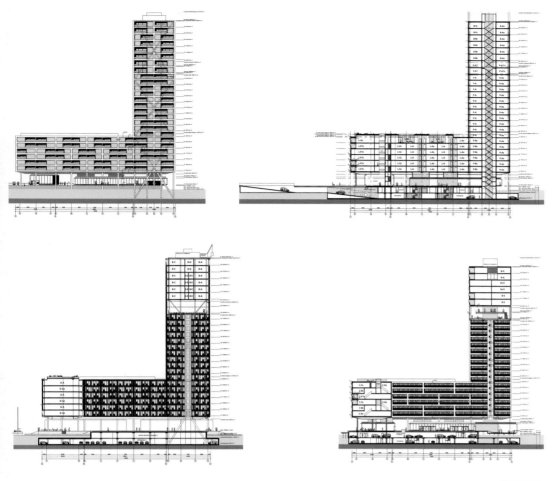

Pontsteiger is the spot where it is possible to realize the unique ambitions of the western shores of the IJ. Living at the head of the Pontsteiger means living in one of the most prominent positions in the IJ.

Because Pontsteiger is elevated and the parking garage is recessed under the water level, the building has a minimal footprint in the river.

The Pontsteiger design has the spectacular view on all sides as its theme. In order to symbolize its unique position in the water city, the building has the form of a gate when seen from a distance. As one approaches, one sees that the Pontsteiger has the form of a chair. The chair has an open side at the back and is elevated 7 m above the jetty. This causes its great volume to seem astonishingly light and airy.

The building has an extremely efficient energy concept, with heat pumps and heat exchangers. This system extracts heat from the deep within the ground. The system will be augmented when necessary by a centralized, highly-efficient heating system. An interesting benefit of heat pump system is their ability to provide cooling in the summer months. Solar collectors and PV panels will be mounted on the roof of the building's high section to help meet the residents' hot water and electricity needs.

Pontsteiger 区是一个最有可能实现 IJ 西海岸壮志雄心的地方。住在 Pontsteiger 的前部意味着住在 IJ 地区最兴旺的位置之一。

由于 Pontsteiger 区是高架形式的，而停车场却隐蔽在水位之下，所以建筑有一小部分是伸进河里的。

让建筑的每一面都有壮观的景观将作为设计主题。为了突显其在水城中的独特位置，建筑从远处看像是一座大门。当靠近它时，会发现建筑是一个椅子般的造型。建筑的背面设计成开放的模式，并且高出码头 7 米，这使得建筑拥有良好的光线和良好的通风。

该建筑融入了高效能源概念，设有热泵和热量交换器。这个系统从地下深处提取热量。同时，在必要的时候通过集中化和高效率的加热系统增强该系统性能。热泵系统的一个益处在于它们能够在夏季提供降温效果，并能持续几个月。太阳能收集器和光伏电池板将被安装在建筑物的屋顶上，以满足居民的热水和电力需求。

Elevation
立面图

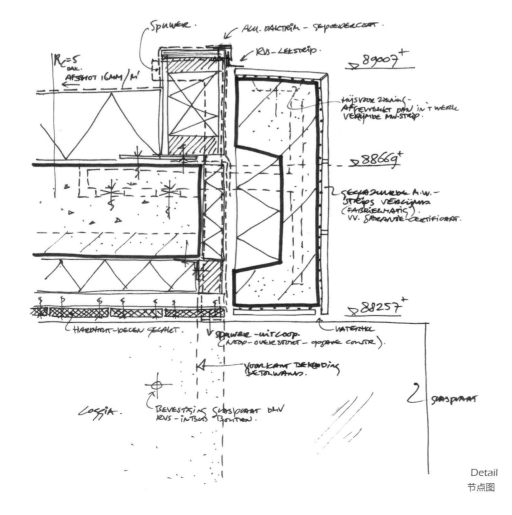

Detail
节点图

SUSTAINABILITY

In addition to geothermal and solar energy, sustainability on architectural shape and façade is also being considered. The large windows allow winter sunshine to penetrate deeply into the apartments, while the deep balconies work as sun shades to keep summer sunshine out. The coolers on northern and interior façades are seldom used to minimize heat loss.

可持续性

建筑不仅采用了地热和太阳能，在造型和立面上都有考虑可持续性。巨大的玻璃窗能在冬季吸收更多的光线到室内，同时深阳台可作为遮阴区防止夏日高阳的照射。而建筑背面和内部的制冷器不必经常使用。

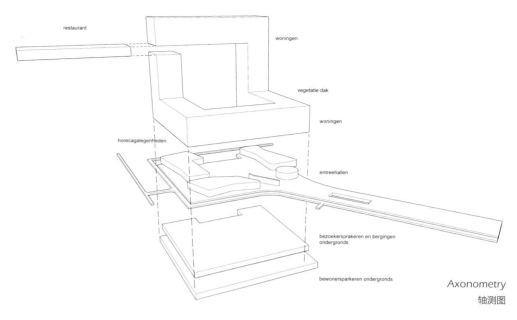

restaurant

woningen

vegetatie dak

woningen

horecagalegenheden

entreehallen

bezoekersprakeren en bergingen ondergronds

bewonersparkeren ondergronds

Axonometry
轴测图

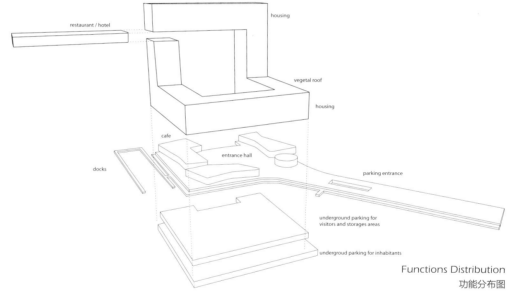

restaurant / hotel

housing

vegetal roof

housing

cafe

entrance hall

docks

parking entrance

underground parking for visitors and storages areas

undergroud parking for inhabitants

Functions Distribution
功能分布图

Model
模型图

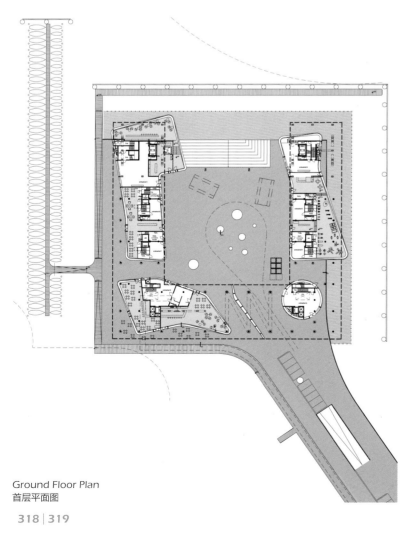

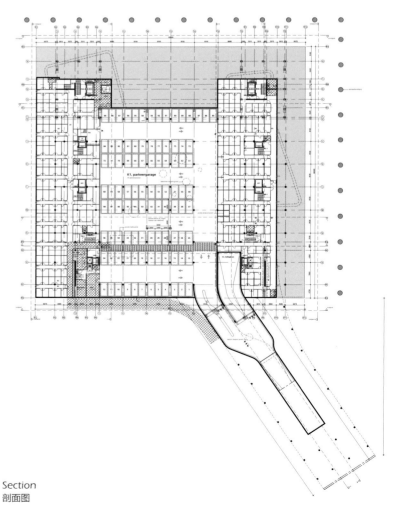

Ground Floor Plan
首层平面图

Section
剖面图

The flat roof of the building's low section will be planted with vegetation, adding green to the views from the apartments above.

To minimize the environmental impact of the building during its life cycle, the façades will not need maintenance. The brickwork uses glue in place of mortar to eliminate maintenance. The exterior window frames are made of aluminum which doesn't require painting. Within the balcony spaces maintenance is possible, which allows the use of other materials as well.

During the interior finishing work, special attention will be paid to health and environmental issues, for example through the use of environmentally friendly paints, PVC-free installations and wood with an FSC certification.

建筑较低部分的平面屋顶上种植绿色植被，为公寓上层的景观增添了一份绿色点缀。

为了减少建筑生命周期内对环境的影响，建筑外墙不需要维护。砖砌建筑物用胶合物替代灰泥以免除过多维护。外部窗框是铝制的，没有油漆。阳台空间可以用其他材料进行维护。

室内的装修工作应特别注意健康和环境问题，例如使用环保涂料，无 PVC 设施和 FSC 认证的木材。

SHAPE 造型

FAÇADE 立面

BALCONY 阳台

SUSTAINABILITY 可持续性

R6, Seoul, Korea

韩国首尔 R6 大厦

Architect: REX
Client: Dreamhub Project Financing Vehicle Co., Ltd.
Location: Seoul, Korea
Site Area: 115,500 m²
Height: 144 m
Renderings: Luxigon, REX
Realization: 2016

设计公司：REX
客户：梦想中心项目融资中介公司
地点：韩国首尔市
占地面积：115 500 平方米
高度：144 米
效果图：Luxigon、REX
完成时间：2016 年

R6 is an urban boutique residence for short-term business people, young urban professionals and foreign residents. Due to the transience of its target users and the short durations during when they are home, R6's unit sizes are small, including 40 m², 50 m² and 60 m² residences, with the majority being 40 m². To meet the trends of its users and compensate for its small unit size, R6 must engender a strong sense of community and its residences must be highly attractive, providing generous views, daylight and cross-ventilation. Maximizing daylight and cross-ventilation are also paramount to provide a highly sustainable residence.

本案是主要供商务人士、城市年轻人以及外国人短期居住、使用的住宅类项目。由于其用户使用的短暂性，项目的单元规模都较小，包括 40 平方米、50 平方米和 60 平方米，其中 40 平方米的居多。为符合用户潮流，弥补较小的单元规模，R6 必须营造比较强的社区归属感，住宅必须有较强的吸引力，提供充足的景观、采光和通风。最大采光和空气对流也是主要因素，以满足高度可持续性。

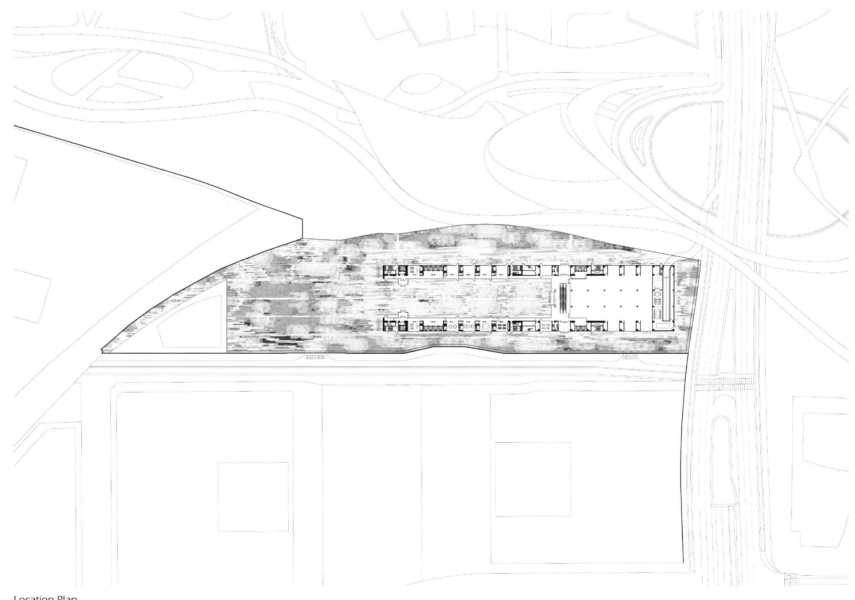

Location Plan
区位示意图

FEATURE 特点分析

FAÇADE

A high-performance façade—composed of frameless IGUs—emphasizes the remarkable exterior views while interior black-out and shade roller blinds control sunlight and glare. The floor to ceiling interior façade—also composed of frameless IGUs and equipped with black-out and shade roller blinds—provides spatial relief and a sense of community while maintaining privacy. The resulting architecture provides views and daylight from both sides and excellent cross-ventilation.

立面

高性能的立面由无框架的中空玻璃单元组成，强调了卓越的外部景观，遮阳卷帘则控制阳光和眩光。室内立面从天花板延伸到地板，同样由无框架的中空玻璃单元组成，并配有遮阳卷帘——作为一种缓和空间的措施，增强社区的感觉，同时保护了隐私。由此产生的体系结构为建筑两边提供了广阔的视野、充足的日光和良好的空气对流。

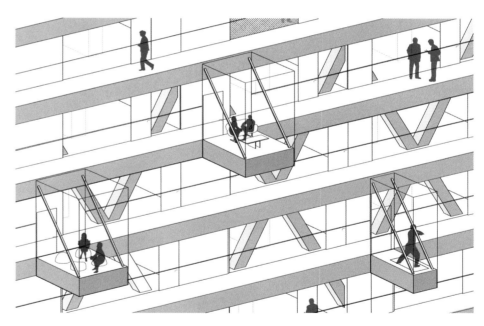

The project looks like a filing cabinet with its drawers open. A hollow centre and large courtyard garden will be revealed at the heart of the 144-metre-high building, which is titled Project R6. A series of compact apartments will overlook this courtyard from within the tower's upper storeys, while shops will surround it at the lower levels.

Block R6 is a narrow parcel bounded by the planned Mountain Park—including Children's Interactive Spray Park, Rail Road Museum, Outdoor Amphitheater, and Yongsan Station Esplanade—and the central park of the planned development Zone B3, adjacent to Hangang-ro. By placing the building to the south of Block R6, all units command great views and the building forms a gateway to YIBD from Hangang-ro.

In a standard housing tower, 40 m2 to 60 m2 units would create poorly dimensioned and oppressive residences, offering constrained views, little daylight, and poor ventilation, and community would be limited to activities at the tower's base.

By pulling layers of the typical housing tower in opposing directions, the small units maintain their size, but are stretched into favorable proportions that provide views and daylight from both sides, excellent cross-ventilation and a strong sense of community through the creation of a central courtyard, roof terraces and conversation/reading/play pods. The stretched layers are strategically positioned to guarantee unobstructed daylight into all units, and to create adequate continuity of the building's primary structure: a concrete-encased steel mega-brace that encircles the courtyard.

项目建筑看起来像一个开着抽屉的文件柜。一个中空的大庭院花园被建在高 144 米的建筑中心，赋予建筑的名称——R6 大厦。一系列紧凑型公寓位于建筑上层，俯视这个花园，同时商店围绕着底层中心而展开排列。

R6 大厦外形狭长，周围环绕着规划中的群山公园，其中有儿童互动喷雾公园、铁路道馆、户外圆形剧场、龙山站滨海艺术中心和中央公园毗邻汉江的 B3 规划发展区。

如果建成普通的塔楼，40 平方米 ~ 60 平方米的居住单元光线会偏差，视野有所局限的视野，整个社区都会显得很压抑，而且活动的地面只能是限于在地面上。

通过拉升建筑对立两面的造型，而小户仍保持原有规模，但将其拉伸至适当的比例后，让阳光可以从两侧照进来，形成良好的空气对流，并通过中央庭院和屋顶平台等的构建创建了一种强烈的社区感。建筑结构作为整体的大型支撑体系。，建筑的结构也创造了许多采光和视野良好的出挑阳台空间。

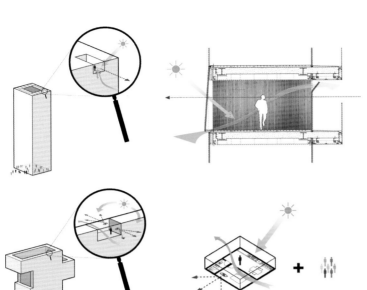

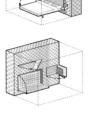

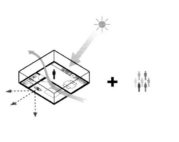

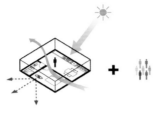

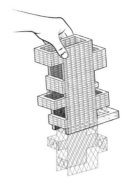

Daylight
日照分析图

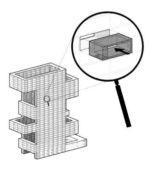

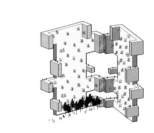

Unit Analysis
户型分析图

Structure Analysis
结构分析图

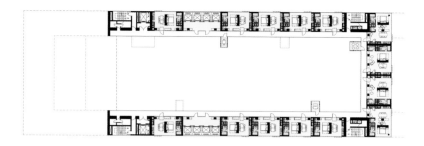

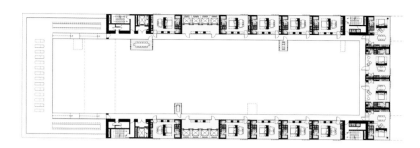

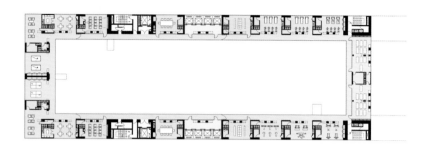

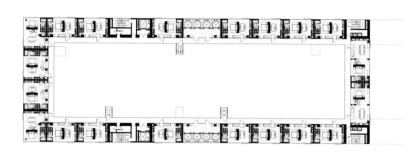

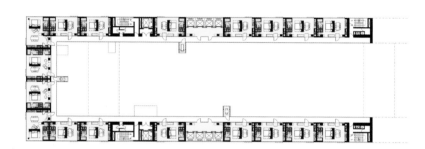

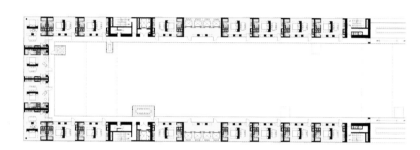

Floor Plan
楼层平面区

The mega-brace supports a shelf-like matrix of walls and floor slabs that define each unit. Into each shelf is inserted a wooden shell containing a bathroom on one side and a kitchen on the other. A movable wall—using standard compact shelving technology—shifts within the unit to define a bedroom (adjacent to the bathroom) or a living room (adjacent to the kitchen). The wall includes a bed, nightstands, couch, television mount, task lights and storage.

这个大型的支撑框架里由架子般的矩阵墙和楼板对每个单元进行划分。在每个架子里插入一个木制结构，内有浴室和厨房。通过标准型紧凑搁置技术创造的可移动墙——可将室内空间随意转换，创造出一个卧室（靠近浴室）或客厅（靠近厨房）。这个墙可容纳一张床、床头柜、沙发、电视、灯和储物柜。

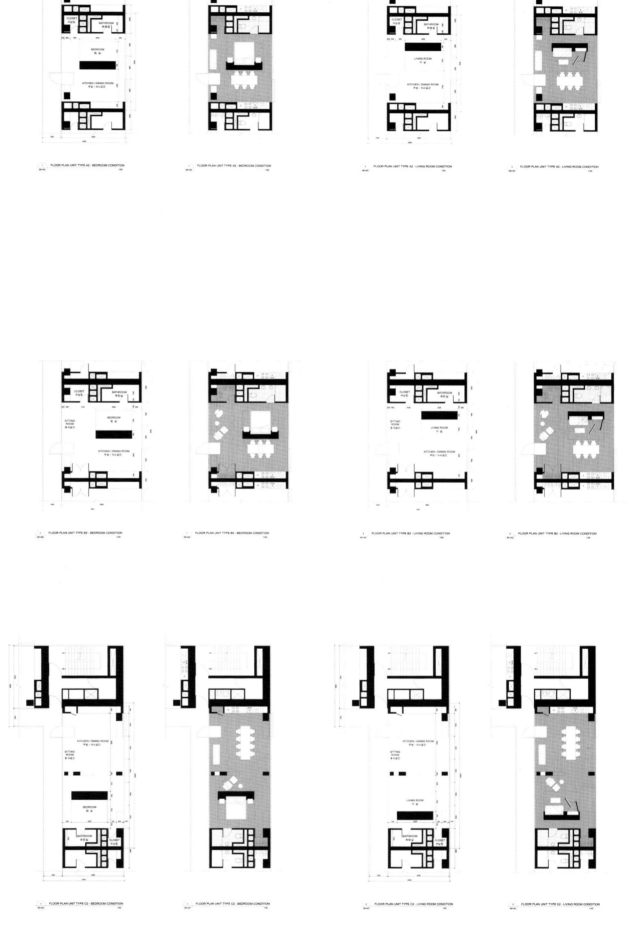

Unit Plan
户型平面图

ROOF 屋顶

FAÇADE 立面

SUSTAINABILITY 可持续性

VIEW 视野

BaiHui Garden, Shanghai, China
中国上海百汇花园

Architect: RTKL Associates Inc.
Client: Shanghai Baihui Real Estate Co., Ltd.
Location: Shanghai, China
Site Area: 24,281m²
Gross Floor Area: 185,000 m²
Photography: Mick Ryan

设计公司：RTKL Associates Inc.
客户：上海百汇房地产开发有限公司
地点：中国上海市
占地面积：24 281 平方米
建筑面积：185 000 平方米
摄影：Mick Ryan

Working with a developer in Shanghai, RTKL designs BaiHui Garden as a premier mixed-use residential development that would appeal to the city's growing demand for a modern urban lifestyle.

由美国 RTKL 建筑设计公司与上海开发商合作的百汇花园是为满足不断增长的城市发展需求，迎合现代城市生活节奏而打造的新一代集多功能于一体的住宅建筑。

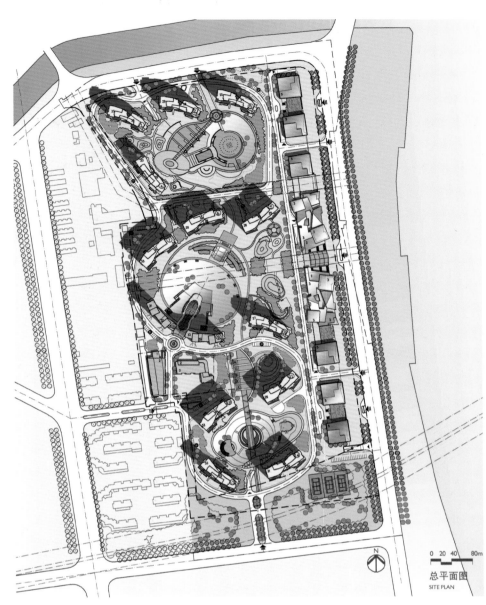

Site Plan
总平面图

Envisioned also as a gateway from central Shanghai to the new master-planned south district, Bai Hui is located in a former industrial section of south Shanghai on the west side of the Huang Pu River. The development occupies 24,281 m² and includes 14 residential and service apartment buildings—ranging from 26 to 33 storeys—40,000 m² of office, retail and restaurant space, a two-storey subterranean clubhouse, and a one-storey subterranean parking structure for 2,500 cars.

The residential and service apartment towers are sited along a north-south undulating line parallel, and echoing with the river. The project's high-rise buildings and lower density will open more space for dynamic landscape. The towers' varied heights will provide the project a distinct landmark profile with its fluid skyline.

The centrally-located clubhouse facilities will offer residents and business travelers an array of recreational amenities. A five-storey retail and restaurant mall will be placed along the edge of the Huang Pu River, forming a riverfront promenade. Two groups of three five-storey office buildings will be sited adjacent to the mall's north and south sides. All of the commercial facilities are connected by underground passageways.

项目坐落于上海南部的前工业区，东临黄浦江，并作为一个通道连接上海市中心区和新规划开发的南部区域，项目占地面积 24 281 平方米，包括 14 栋从 26 层到 33 层的住宅楼和服务公寓建筑，40 000 平方米空间设置为办公区、零售区和餐饮区，一个两层的地下俱乐部，和一个可容纳 2 500 辆车的单层地下停车场。

住宅和服务公寓沿江而建，南北起伏，呼应江水。项目的高层楼宇和低密度为动态景观开放出更多空间。建筑的高度变化营造了一种独特的轮廓，并以其流体的天际线形式赋予了建筑别样的标志性意义。

建筑的中央位置将设置一系列娱乐休闲设施，为居民和商旅者提供服务。一栋 5 层楼高的零售、餐饮、购物中心沿江设置，形成了一条江畔步行街。六座 5 层高的办公大楼分两组分别置于购物中心的南北两侧。同时，地下通道的设置将所有的商业设施紧密连接起来。

Site Elevation
总体立面图

FEATURE 特点分析

FAÇADE

The bulge balconies look like the wave flowing through the building façade, echoing with the view of Huang Pu River, and bringing a sense of dynamism and vitality to the façade of the building.

立面

凸出的阳台如波浪般流动在建筑的立面，呼应了黄浦江的江景，为建筑的立面带来动感和活力。

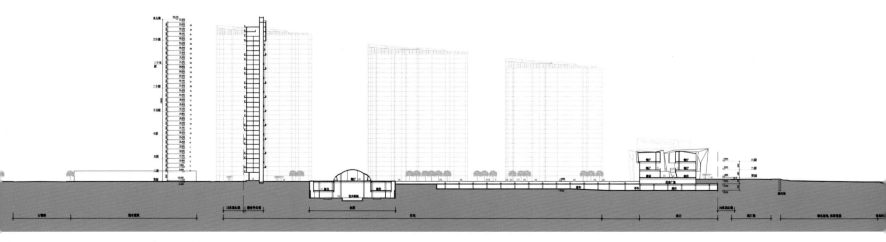

0   15   30   60m

Site Section
总体剖面图

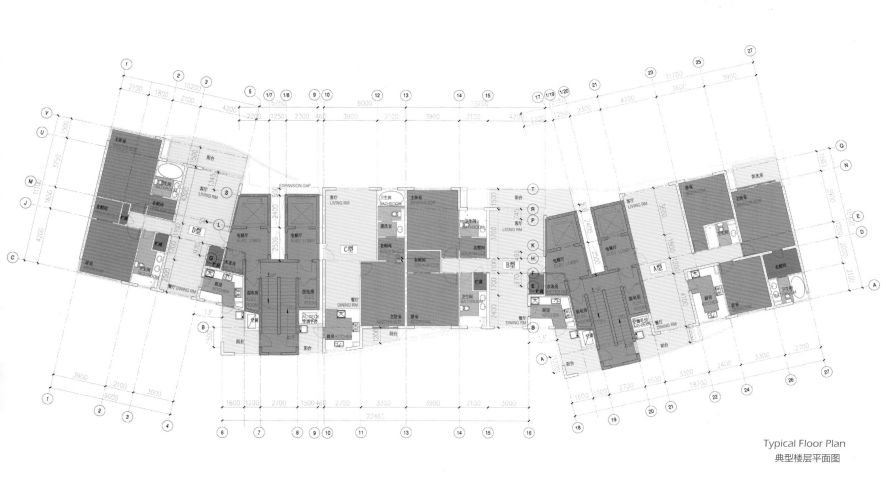

Typical Floor Plan
典型楼层平面图

MATERIAL 材料
FAÇADE 立面
BALCONY 阳台
STRUCTURE 结构

Sugar Dock, Sydney, Australia
澳大利亚悉尼蜜糖码头公寓

Architect: Francis -Jones Morehen Thorp
Client: Lend Lease
Location: Sydney, NSW, Australia
Gross Floor Area: 16,506 m²
Photography: Andrew Chung, John Gollings

设计公司：Francis -Jones Morehen Thorp
客户：Lend Lease
地点：澳大利亚新南威尔士州悉尼市
建筑面积：16 506 平方米
摄影：Andrew Chung、John Gollings

The Sugar Dock is located at Jacksons Landing, an area which has been subject to considerable revitalization over recent years, with a number of new residential apartment buildings developed by Lend Lease and public open spaces being constructed in the area immediately surrounding the subject site. Nearing completion, the master-planned site has developed a strong residential community and valued address.

"蜜糖码头"住宅楼位于澳大利亚近年来发展迅猛的 Jacksons Landing 地区，开发商 Lend Lease 已经在该地区开发了一系列住宅公寓，开阔的公共空间直接建造在主要工地周围。项目接近完成时，这个总规划已经发展成为庞大的住宅区域和最具价值地区。

Sugar Dock provides 132 residential units associated with the communal open space and underground car parking. The design of the development seeks to complement and responds to its urban context characterized by its relationship to large scale elements such as the Harbor, the working dockyards, silos and the Anzac Bridge as well as the immediate topography and the streetscape of Jacksons Landing. The tower consists of a series of glass residential apartments enclosed and anchored by two precast shells, combining the visual language of the new development with that of the old Pyrmont. Within the glass form, apartments are turned away from their primary boundary towards the view and optimal orientation.

A podium element grounds the tower's masonry walls and reinforces the sandstone cliff-like Bowman Street alignment. Apartments along Bowman Street have been articulated to read as a series of houses inserted into the cliff face. They have an intimate character with their connection to the park in Bowman Street and are restricted in height allowing clearer views through site views from the Distillery Hill and the communal open space.

"蜜糖码头"提供132个住宅单元，还有相连的公共开阔空间及地下停车空间。设计旨在呼应和弥补场地内的城市环境，这个城市环境是由与大规模元素，如海港、造船所等之间的关系来定义的。大厦包含一系列玻璃住宅公寓，由两个预制的壳体包裹和定义，设计将新建筑的视觉语言与附近老建筑的语言相结合，完美呼应。这样的玻璃造型，让公寓拥有尽览美景的绝佳区位方向。

底层的基座采用石墙包围，与鲍曼大街上排成直线的砂岩状建筑相得益彰。公寓沿着鲍曼大街而建，看上去好像是崖壁上嵌入的房屋。楼宇与鲍曼大街上的公园紧密相连，同时，严格的高度限制使人们可以从住宅楼的公共空间和Disillery山更清晰地看到附近公园的景色。

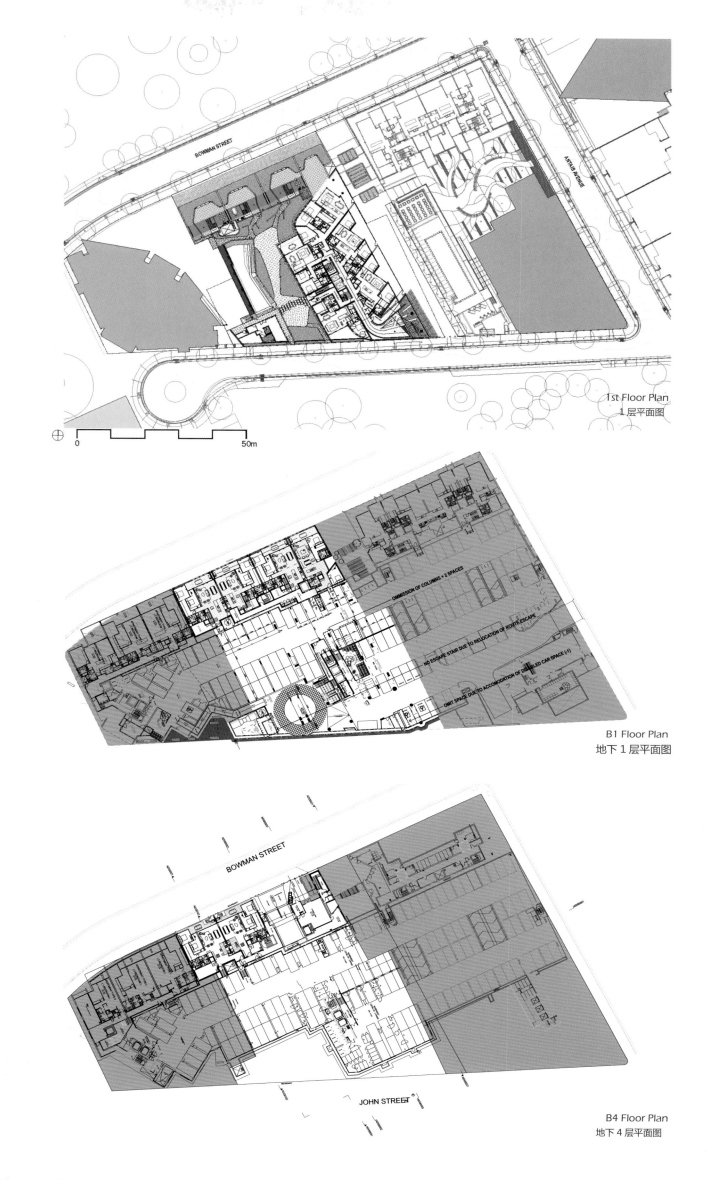

1st Floor Plan
1层平面图

0　　　　　　　　　　50m

B1 Floor Plan
地下1层平面图

B4 Floor Plan
地下4层平面图

FAÇADE

Building façade expresses the relationship with the surrounding environment by using the materials of stone, precast cement, glass and other different words. Stone continues the culture of old street; precast slab contrasts with the surrounding environment, highlighting the personality, and the element of glass brings good sight.

立面

建筑立面使用石材、预制板水泥板、玻璃等不同的材料语言来表达与周围环境的关系。石材延续了老街道的文化；预制水泥板与周围环境形成对比，突出了个性；玻璃部分带来了良好的视线。

The material, chosen for the shells, is vertically articulated light-coloured precast concrete providing a relationship with the prominent sandstone outcrops of the Distillery Hill development. The elevations have been developed to correspond with more private functions and to provide protection from afternoon sun. To the northeast, a metal/glazed balustrade system has been developed to utilize daytime reflections to provide a surface to the building at the balcony line, suppressing the slab edge.

This system continues to the parapet to cap the building and provide environmentally controlled enclosure to the penthouse open spaces. Behind the balustrade is a series of fritted glass panels, arranging to create a greater sense of privacy and to provide a "screen behind a screen" effect which operates both at the individual apartment scale and the scale of the whole elevation.

The metal/glazed system provides a lightness to the building's material character appropriate to its height and heightenes environmental screening to the recessed balconies.

在立面材料的选择上，定制的垂直铰接浅色混凝土和 Distillery 山项目突出的砂岩基石形成鲜明的对比。这样的立面是为了满足个人需求而设置的，同时避免下午阳光的直射。东北方向，金属/玻璃栏杆系统的运用充分利用了白天的反射效果，为建筑的阳台边缘提供了一种新立面，反过来压制了天板边缘。

栏杆系统覆盖了建筑整体，并作为顶层开放空间环境控制的护栏。栏杆后面是一系列多孔玻璃面板，体现了一种强烈的隐私感，并创造了一种"屏风后的屏风"的效果，同时运用在个人公寓和整栋建筑上。

金属/玻璃系统为建筑材质提供一种轻盈感以适应建筑的高度，并且为凹形阳台带来了丰富的景观。

4th Floor Plan
4 层平面图

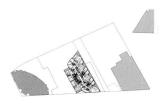

SHAPE 造型

FAÇADE 立面

SUSTAINABILITY 可持续性

STRUCTURE 结构

The Residencia, Macau, China

中国澳门君悦湾

Architect: P&T Group
Client: San You Development Co., Ltd.
Location: Macau, China
Site Area: 13,900 m²
Gross Floor Area: 231,000 m²

设计公司：P&T 集团
客户：三友发展有限公司
地点：中国澳门特别行政区
占地面积：13 900 平方米
建筑面积：231 000 平方米

As the Hong Kong-Zhuhai-Macau Bridge construction is approved, Oriental Pearl District soon becomes the next generation of Macau's potential growth area. The new project—the Residencia Macau-is located here, attracting the eye of investors with its charming sea-view, luxurious clubhouse, lots and other advantages to create the extremes of luxury properties in Macau.

随着港珠澳大桥基建拍板，澳门东方明珠区旋即成为新一代澳门潜力优厚的发展区域，而坐落该区的焦点新盘——澳门君悦湾，以其迷人海景、华贵的会所、优越的地段等优势，将成为澳门豪宅之最，吸引了不少投资者的目光。

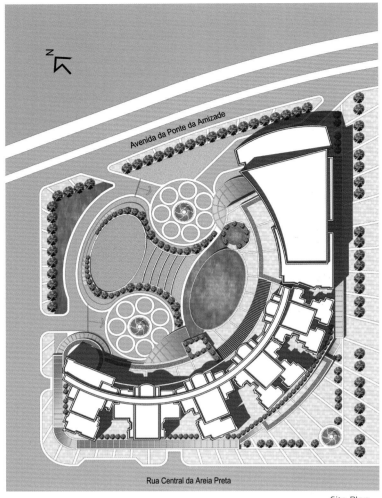

Site Plan
总平面图

Located in the new district of luxury residential district comparable to West Kowloon, Hong Kong, most of the three-bedroom units face the sea so as to embrace a prosperous future, overlooking the coastline of Zhuhai and Waterfront Park nearby. Community facilities are so well-equipped. There are also infinite pools and gymnasium in the club. The building with elegant materials and luxury design will win the favor of residents. Located on the Avenida de Amizade, the future Macau landing site of the Hong Kong-Zhuhai-Macau Bridge, the Residencia Macau is a high-end residential development occupying a total land area of 13,900 m² fronted by the full panoramic sea-view of the South China Sea.
Completed in 2009, the project's private landscaped seaside garden commands tranquility by day, and provides a dramatic contrast with that of the exciting cityscape view of nearby Zhuhai by night.

Towering between 45 and 49 storeys, the Residencia Macau is organized like a curved horn on plan. It is a mixed-use development which comprises seven linked blocks strung along the curve, featuring a hotel, a serviced apartment block and five residential towers.

The scheme is anchored by a banqueting hall in the basement floor and a six-storey podium of clubhouse facilities, car parking, retail outlets and F&B outlets in the hotel, together with the 5,574.18 m². Landscaped garden at ground level provides a recreational area for residents and visitors.

The vision for the project is the building of a structure with a refreshing contemporary outlook that will respond to the new wave of market diversity and sophistication.

建筑坐落于媲美香港西九龙的豪华住宅新地段，绝大部分的三房单位面海，占尽天时地利，繁华前景无限，远眺珠海海岸线，近揽海滨公园。小区设施极为完善，生活所需无一或缺。会所设计有无边际泳池、健身馆等。建筑用料考究，设计豪华，必将赢得住户的青睐。项目位于友谊桥大马路，临近未来港珠澳大桥的澳门落脚点，占地 13 900 平方米，是澳门一线海景楼盘，坐拥中国南海完整的景观。

项目于 2009 年竣工，其私人滨海花园在白天显得格外宁静，与珠海夜间璀璨的海景形成了鲜明的对比。

建筑从 45 层到 49 层不等，呈弯曲的角状组织。建筑是一个多功能的综合性项目，包括七个呈曲线状展开的连续区域：旅馆、酒店式公寓和五栋住宅楼。

一个大型宴会厅设置在地下层，而六层的墩座部分包括俱乐部会所、停车场、零售点和餐饮连锁酒店。景观花园占地 5 574.18 平方米位于地面层，供居民和游客休闲之用。

建筑结构给人耳目一新的感觉，以应对新一轮市场多样性和复杂的挑战。

FEATURE 特点分析

FAÇADE

To reinforce a continuous surface treatment, high-grade Belgian reflective glass is used throughout the curtain wall. The glass matches with the refined and elegant architectural curve to create a multi-dimensional visual effect which differentiates the development from other nearby residential projects.

立面

为了加强建筑表面的连续性，高档比利时反光玻璃幕墙覆盖了整栋建筑。玻璃搭配高雅脱俗的建筑曲线创建出一种多维的视觉效果，这将建筑与周围的住宅项目明显地区分开来。

Typical Floor Plan
典型楼层平面图

SHAPE 造型

FAÇADE 立面

STRUCTURE 结构

SUSTAINABILITY 可持续性

The Verv @ RV

Verv @ RV 公寓大厦

Architect: ONG&ONG Pte Ltd
Client: Heritage @ River Valley Pte Ltd
Location: Singapore
Site Area: 955.8 m²
Gross Floor Area: 2,676. 24 m²
Photography: ONG&ONG Pte Ltd

设计公司：ONG&ONG Pte Ltd
客户：Heritage @ River Valley Pte Ltd
地点：新加坡
占地面积：955.8 平方米
建筑面积：2 676. 24 平方米
摄影：ONG&ONG Pte Ltd

The Verv@RV is located in the heart of River Valley Road, Prime District 9, Singapore, next to Orchard Road, and within walking distance to Sommerset and Orchard MRT as well as Orchard shopping belt and Great World City. It's close to Marina Bay New Downtown, central business district, Clarke Quay and Boat Quay.

本项目位于新加坡第九区的河谷路中心区，临近乌节路，而到萨默塞特、乌节捷运站、乌节购物区和伟大世界城市也只需步行就可到达。同时，它临近滨海湾新市中心、中央商务区、克拉克码头和驳船码头。

Rendering
效果图

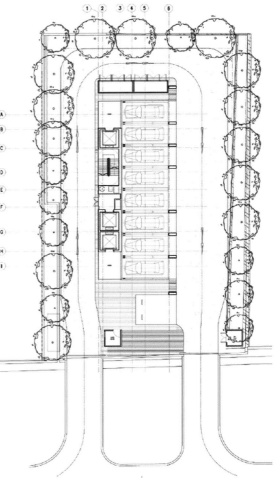

Communal facilities like a 20-m-long lap pool with Jacuzzi and an open gymnasium are thoughtfully elevated and housed on the second storey. To circumvent the limited land size, a mechanized parking system is installed enabling passengers to be dropped off at the private lift lobby.

A logo "Verv@RV" is created for the branding of the building and it is used as the overall design concept of the façade giving it a distinct identity. A secondary mark is distilled from the logo and this pattern spans the full height aluminium trellis forming the building's backbone and is also in the horizontal entrance lattice.

This pattern follows through consistently and is applied to the aluminium perforated screen that covers the mechanised parking system and the private lifts. Finally, the use of colour shading of the same motif is created on the entrance driveway, creating playful shadows when the sun shines on it.

The design concept is inspired from its brand which is also the origin to create its shape and configuration.

Garage
车库

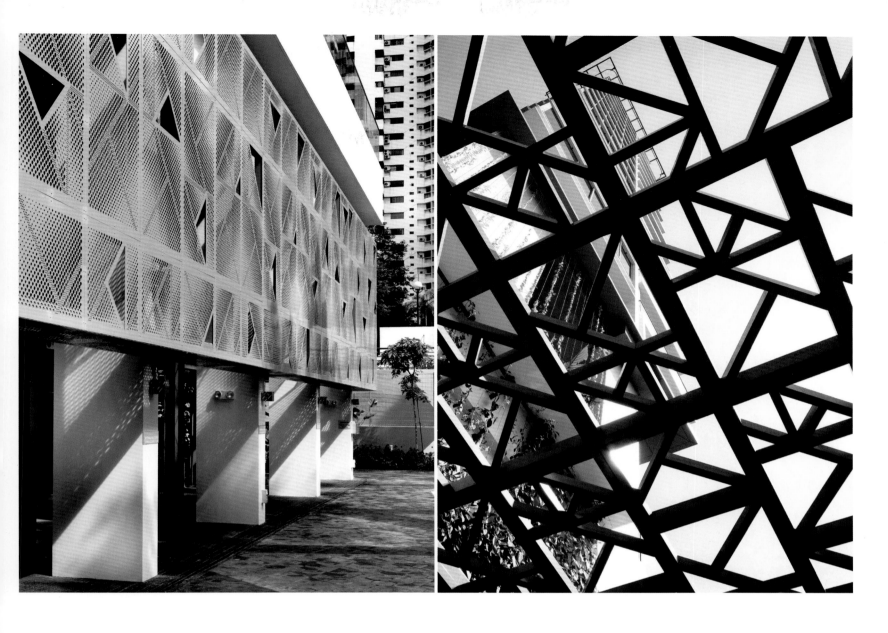

FEATURE 特点分析

FAÇADE

The grids of the façades not only protect privacy, but are also used as a shield for solar protection.

立面

建筑网格的立面设计不仅可以保护个人隐私，也可以作为一种屏蔽，起到良好的遮阳效果。

公共设施被提升设置在建筑的第二层，如直径 20 米的小型健身游泳池和开放体育馆。为避免占用有限的空间，设计师安装了机械停车系统，使得乘客可以乘私人电梯进出。

Verv@RV，这个商标源自于建筑品牌，并作为立面的总设计概念赋予它与众不同的特质。一个副标志从商标中被提炼出来，这个模式从垂直方向上横跨铝制格子构架，形成了建筑的主干，同时在水平方向上，融入到格子的入口。

这种连续性的模式同时被应用于铝制多孔屏风，覆盖了机械停车系统和私人电梯。最后，在车道入口处，同样主题的阴阳色的运用在阳光的照耀下创造出有趣的光影效果。

项目设计理念源于品牌，这也正是其造型、配置的基础。

SUSTAINABILITY 可持续性

FAÇADE 立面

STRUCTURE 结构

GREENERY 绿色

Helios Residences, Singapore

新加坡嘉旭阁

Architect: Guida Moseley Brown Architects

Location: Singapore

设计公司：Guida Moseley Brown Architects

地点：新加坡

Helios Residences is located in one of Singapore's most densely developed residential areas, a precinct including towers and compactly positioned terrace houses, all within close walking distance of the linear retail and entertainment heart of the city, Orchard Road. The roughly Y-shaped site has a 12 m topographical variation with the primary vehicular access at the uppermost point at Cairnhill Circle. Three closely interlinked towers "zig-zag" into the "Y" shape, and the first occupied floor is one level above Cairnhill Circle, leaving the site to step down with terraces, gardens and courts, with car parking structures below these various amenities.

嘉旭阁位于新加坡的高密度住宅区之一，这里高楼林立，排屋鳞次栉比，城市的商业零售和娱乐中心——乌节路就近在咫尺。场地呈"Y"型，其中有一段12米长且地形变化不同，主车道在最顶点的经禧圈。本项目的三栋大楼在"Y"字部分相互曲折连接，首层部分比经禧圈高一个楼层，以露台、花园、球场作为过渡连接，停车场则设置在这些设施的下方。

FEATURE 特点分析

FAÇADE

The window glazing is high-performance LOW-E type and represents less than 50% of the total elevations. The spandrel of the curtainwall construction is low-heat retention due to glass-block, air space, back panel assembly, which together with the glass makes a high-performance external envelope.

立面

玻璃幕墙采用高性能的低辐射材料，占据了总立面面积的50%。玻璃幕墙之间的拱肩结构具有低热特性，玻璃体量、空中空间、后面板组件和玻璃立面一起组成了高性能的外部立面结构。

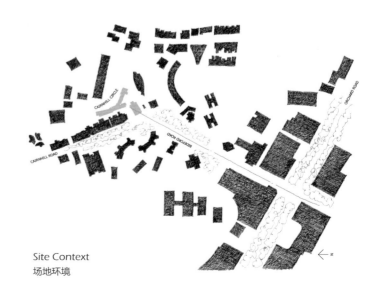

Site Context
场地环境

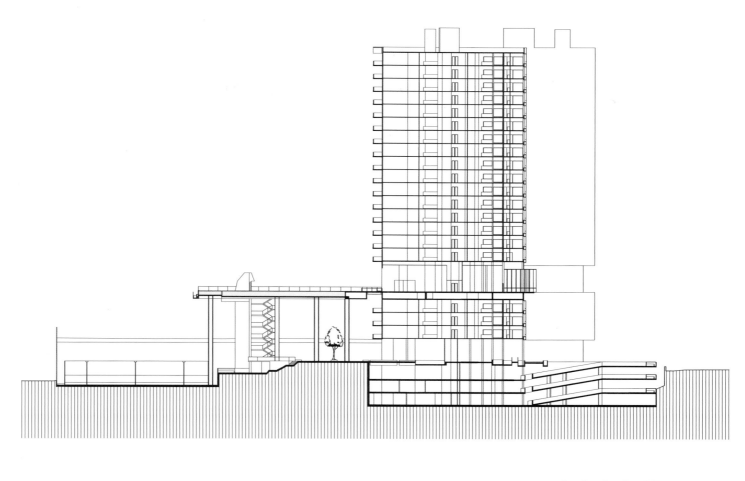

0 5 10 15 20 METRES

Section
剖面图

Because the building is positioned in the city on a long axial view from the significant intersection of Orchard Road and Bideford Road, it has been designed with architectural form and lighting to act as an urban marker. The lighting integrated into the curtainwall assembly is comprised of low-energy LED lighting strips, and is set to be at selected and limited hours. The lighting is accessible through swing panels of glass block within each spandrel panel. The low level of illumination creates a subtle distinction of the tower at the evening hours.

These architectural elements are positioned at the storey 4, a double-height sky terrace providing various amenities for residents: gym, child's play areas and pools, changing rooms and a number of landscaped social activity places. As well, at the western perimeter of the towers, an independently supported swimming pool deck extends from this level, 15 m above a significant garden.

由于建筑位于城市的视轴之中——乌节路和必德福路的交叉路，所以设计师希望其建筑形式和照明能成为城市的标志。照明元素被融入到玻璃幕墙中，由LED照明带组成，可选择性和分时段发光。每个拱肩上的摆动的玻璃面板让灯光更为炫目，低层的照明则在夜间营造出微妙的氛围。

不同的建筑元素都设置在第4层，有双高的空中露台，儿童游乐场和泳池、健身房、更衣室及各种环境优雅的社交场所能为居民提供各种服务。同时，西侧的大楼有独立的泳池，其甲板从15米高的上空向外延伸，下方是一座美丽的花园。

The elevated swimming pool has allowed an extensive garden at ground level, which emanates from the vehicular arrival drop-off, and is developed with a highly integrated landscape of trees, low plantings and extensive bodies of water. An all-glass block multi-purpose room opens onto one of the stepped terraces of this secluded garden and a 90-m-long, 5.5-m-high curved water wall tempers the environment, masks vehicle noise, defines the change in elevation from the upper road level, and anchors the various landscape spaces.

The design has been executed in an effort to maximize the benefits of the site conditions and the plan creates two linked towers at the north-facing upper-level Cairnhill Circle, with a third tower fronting onto Cairnhill Road and terminating a visual axis from Orchard Road. This later tower has, as its first level, the storey 4 sky terrace supported on 25-m columns, with the space between a veil of vines growing on cables tensioned between the sky terrace and the ground. This three-dimensional green screen is intended to contribute to the green character of large and mature street trees along Cairnhill Road.

首层的大花园则是为高空泳池做铺垫，也是落客区的入口，这里树木花草茂盛，还有丰富水景。一个全玻璃里面的多功能室朝这个隐蔽公园其中一个阶梯式平台开放；一座 90 米长、5.5 米高的弧形水墙不仅让环境更显幽静，还能缓和车辆噪音，让上方道路变得美丽的同时也提供了不同的景观空间。

设计最大化地利用场地条件，将两栋大楼布置在朝北的经禧圈上层，第 3 栋大楼则是面向经禧路，挡住了延伸到乌节路的视轴。第 3 栋大楼的首层和第 4 层的空中露台都由 25 米高的梁柱支撑，支柱之间的空间有葡萄藤沿着空中露台和首层延伸出的电缆而生长。这片立体绿屏将成为经禧路上许多大型成熟树木的延续。

The structural systems are efficient and repetitive to minimize the use of structural materials as much as possible (energy reductions in production, usage and transportation); the plans, developed for effective ventilation and the overall coverage of the occupied spaces, are a high-performance curtainwall system. The building's location on the site takes advantage of shading from adjacent structures and the design of an overall green environment lessens heat build-up on the site. Additionally, rainwater is collected in tanks for reuse in irrigation.

The architecture is simple to act as a backdrop for these various activities and places, allowing focus on the landscape and water elements, and the activities of the residents. Nevertheless the design of the towers with the linear composition of the blocks, spandrels of glass block, the varyingly placed balconies, the penthouse lifts and the ever-present greenery create a distinct architecture of memorable character and a sense of place designed to age gracefully.

建筑内各处都采用了高效的结构系统，以尽量减少结构材料的使用（降低生产、使用和交通方面的能量消耗）；为达到通风顺畅和整体遮阳的效果，采用高性能玻璃幕墙系统。建筑所在的位置能巧妙地利用相邻建筑结构的遮蔽，加上整体绿色环境的设计，能让整个场地减少大量热辐射。此外雨水也被收集到水箱中做灌溉循环利用。

建筑结构简洁，可作为各种活动的举办场地，优美的景观和水景也为居民提供休闲场所。各大楼的线性组合、玻璃拱肩、各种各样的阳台、复式住宅电梯、无处不在的绿化，都赋予了建筑令人难忘的特征，也是享受生活的绝佳场所。

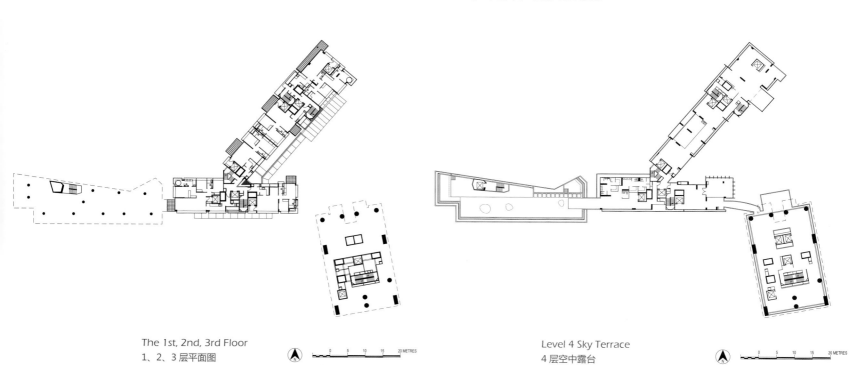

The 1st, 2nd, 3rd Floor
1、2、3层平面图

Level 4 Sky Terrace
4层空中露台

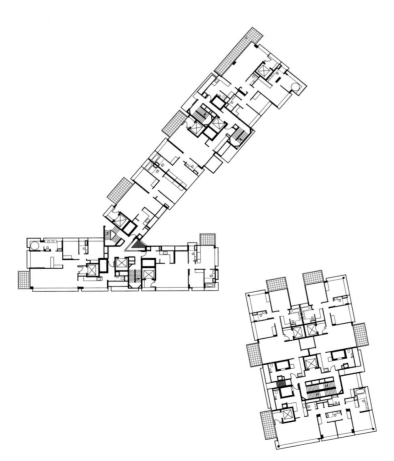

The 6th, 8th, 10th, 12th, 14th, 16th, 18th Floor
6、8、10、12、14、16、18 层平面图

0   5   10   15   20 METRES

SHAPE 造型

FAÇADE 立面

BALCONY 阳台

MATERIAL 材料

Draycott Park Residential Complex

Draycott 公园住宅综合体

Architect: Guida Moseley Brown Architects
Location: Singapore
Floors: 24

设计公司：Guida Moseley Brown Architects
地点：新加坡
楼层：24

The design approach to the Draycott Park Residential Complex is one that is based simultaneously on a respect for the past and a positive attitude to our contemporary time.

Draycott公园住宅综合体的设计手法既体现了对过去的尊敬，又迎合了当今时代的需要。

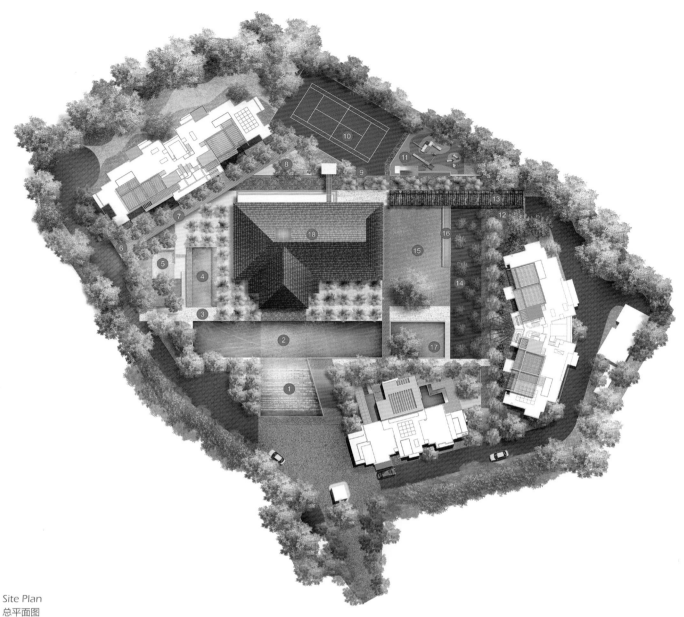

1. Entrance Waterfeature
2. 50m Lap Pool
3. Pool Deck
4. Childrens' Pool
5. Jacuzzi Pool
6. Drop-off Pavillion
7. Covered Linkway
8. Fitness Area
9. Green Featurewall
10. Tennis Court
11. Children's Playground
12. BBQ Areas
13. Garden Trellis
14. Timber Deck / Planter
15. Multi-purpopse Lawn
16. Water Trough Waterfeature
17. Reflective Pool
18. Clubhouse
   a) Party Room with Annex pantry
   b) Children's playroom
   c) Internet Salon
   d) Theatre Lounge
   e) Book Lobby
   f) Reading Foyer
   g) Male/Female Powder cum Changing Rooms
   h) Steam Room
   i) Sauna Room
   j) Locker facilities
   k) Day Bed Terrace
   l) Concierge desk
   m) Gymnasium
   n) Yoga / Pilate Room
   o) Onsen Bath cum Powder Room
   p) Games Room
   q) Billard Room
   r) Wine Cellar
   s) Wine Tasting Lounge
   t) Cigar cum wine tasting terrace

Site Plan
总平面图

FEATURE 特点分析

FAÇADE

Every unit has corner windows providing dual exposure. The loft units are of two-storey height, and each living room opens out onto balconies overlooking the central gardens. Special water features, panels of glass block, moveable metal louvres and dichroic glass provide an environment of constantly changing character and of considerable visual interest. Every elevation is different, yet participates in the creation of a carefully balanced unique sculptural form.

立面

每个单元设置有角落窗户以提供双重照明。阁楼有两层楼高，每个客厅配置有阳台，可以俯瞰中央花园。特殊的水景、玻璃面板、可移动金属百叶窗、双色玻璃创造出不断变化的环境和丰富的视觉体验。虽然各立面各有不同，却又能组成和谐又独特的雕塑造型。

Within a complex of three 24-storey buildings, the design provides two tower types: two large towers of four-bedroom units in a reserved classic modernist approach to massing, composition and detail, and one point-block tower of two-bedroom units in a freer and more expansive, sculptural character.

Both building types are developed with large-scale architectural elements that reduce the overall scale of the buildings, creating sub-scales of form appropriate to the creation of a setting for the centrally placed two-storey historic house which has been renovated and used as the club for the condominium. Importantly and properly, the central landscape design develops its initial planning and character from the formality of the historic building, and integrates each of the tower buildings into a comprehensive design of green landscape, active and passive bodies of water, and well-formed open spaces. Also important is the fact that the landscape has been conceived together with the architecture and entry canopies, covered walkways, garden walls, trellises and similar devices to create a human scale that links with the historic house and to create a comfortable atmosphere.

三栋建筑高度为 24 层，有两种设计方案：两栋高大的塔楼在造型、组合和细节方面保留了经典现代主义方法，内部是四居室单元；另一栋大楼的设计更为自由、宽敞，雕塑感更强，内部是两居室单元。

两种设计方案采用了大规模的建筑元素，以缩小整体建筑规模来形成适当的区域空间。这样做是考虑到位于中心的一栋具有历史意义的两层房屋，它已被重新翻新作为住宅区的俱乐部。更重要的是，中央景观的规划和特征设计从历史性建筑的禁锢中解脱出来，与每一栋建筑融合，形成统一的有绿色景观和流动水景的开放公共空间。另外，景观的设计构思将建筑、入口遮阳棚、人行走廊、花园墙壁、凉亭和其他类似设施融合在一起，再与历史意义浓厚的房子结合起来，打造了一个舒适的氛围和尺度宜人的环境。

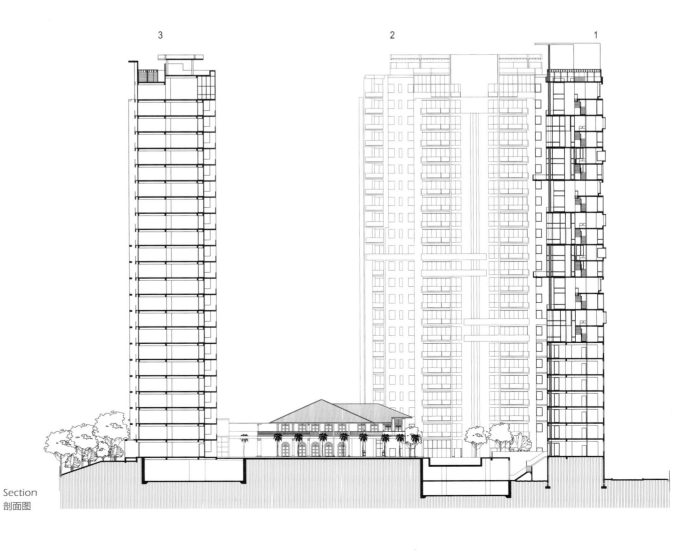

Section
剖面图

The 20th Floor
20 层平面图

The 19th Floor
19 层平面图

The 18th Floor
18 层平面图

The 17th Floor
17 层平面图

The 16th Floor
16 层平面图

The 15th Floor
15 层平面图

The 3rd Floor
3 层平面图

Tower1 Lever 15th Plan
1 号楼 15 层平面图

Tower1 Lever 16th Plan
1 号楼 16 层平面图

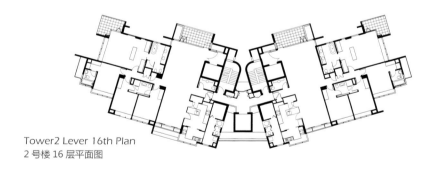

Tower2 Lever 16th Plan
2 号楼 16 层平面图

Tower3 Lever 16th Plan
3 号楼 16 层平面图

The single tower of two-bedroom units provides a wide range of different unit types, configured to allow for considerable choices and lifestyles. The plan incorporates three types of single-storey units and eight different loft units. The single-storey units are at the lower levels and are close to the central garden, and the lofts are varied in location to create the sculptural form of the tower. The loft units provide numerous opportunities to create unique living environment, which not only incorporate many of the architectural elements of the base units, but also can be developed with creative interior design.

The two different types of towers utilize similar materials so as to have a harmonious relationship, yet they use them differently. The buildings incorporate white walls with areas of colour, and areas of silver aluminium that will reflect the skylight, adding this natural element into the design. Stone is employed at appropriate entry areas.

内部是两居室户型的大楼提供多种单元类型，可以满足不同的需求和生活方式。户型规划包含了三种单层单元和八种阁楼单元。单层单元位于低层，靠近中央花园，阁楼单元设置在不同的位置以突出建筑的雕塑感。阁楼单元能创造独特生活环境，它不仅将各单元的诸多元素融合到一起，还能提供创新的室内设计。

另两栋大楼类型各不相同，虽然都采用了类似的材料，以构建一种和谐的关系，但使用方式却又不相同。设计融入了白色底墙、区域性色彩、反光银铝板等自然元素，而石头则运用到入口的适当区域。

ACKNOWLEDGEMENTS

鸣谢

The Architectural Design & Research Institute of Guangdong Province
Collection House for Scientists in Guangzhou Science City, Guangzhou, China

BIG, Westbank, Dialog, Cobalt, PFS, Buro Happold, Glotman Simpson, James Cheng
Beach and Howe St, Vancouver, Canada

WSP Architects
1 Unit • 100 Families • 10,000 Residents, Shenzhen—Happy Town

RSP Architects Planners & Engineers (Pte) Ltd
The Pinnacle @ Duxton, Singapore
The Inerlace, Singapore

Arc Studio Architecture + Urbanism
The Pinnacle @ Duxton, Singapore

AXS Satow Inc.
Collection House for Scientists in Guangzhou Science City, Guangzhou, China

Renzo Piano Building Workshop
London Bridge Tower(The Shard), London, UK

Adamson Associates
London Bridge Tower(The Shard), London, UK

The Architecture Design & Research Institute of Guangdong Province Branch of Shenzhen
Wu Zi Apartment, Shenzhen, China

Zaha Hadid Archtiects
Bratislava Culenova New City Centre
Dorobanti Tower, Bucharest, Romania

Jiang Architects & Engineers
Sanya Phoenix Island, Sanya, China
Changjing Zhige ABC1 Lands Mixed-use Development in Wuhu, China

Steffian Bradley Architects
Wu Zi Apartment, Shenzhen, China

SOM
WKL Hotel and Residences, Kuala Lumpur, Malaysia

dwp | design worldwide partnership
HQ by Sansiri, Bangkok, Thailand

UNStudio
V on Shenton, Singapore
The Scotts Tower, Singapore

Studio Daniel Libeskind
Reflections at Keppel Bay, Singapore

OMA
The Interlace, Singapore

P&T Group
M Ladprao, Bangkok, Thailand
The Residencia, Macau, China

Heatherwick Studio
Sheung Wan Hotel, Hong Kong, China

Arquitectonica
Infinity, San Francisco, USA

Zeidler Partnership Architects
The Helix

HOK
Baku Flame Towers

Arup
Taipei Bade Urban Renewal Residence, Taiwan, China

Tabanlioglu Architects
Levent Loft II, Istanbul, Turkey

Dominique Perrault Architecture
"LA LIBERTE" Housing and Office Building

SOMA
Aura

Gokhan Avcioglu & GAD
YDA

C. F. Møller Architects
Continental Tower, Stockholm, Sweden

Asymptote Architecture
Velo Towers, Seoul, South Korea

Foster + Partners
South Beach, Singapore

Aedas
South Beach, Singapore

Contexture Architects
Essentials Dream Town Qingdao – Amsterdam Block

RTKL Associates Inc.
Cosmopolitan, Puerto Rico
BaiHui Garden, Shanghai, China

DBI Design
Etihad Towers, Abu Dhabi, UAE

MAD
Sanya Phoenix Island, Sanya, China

Francis -Jones Morehen Thorp
Sugar Dock, Sydney, Australia

REX
R6, Seoul, Korea

SPARK
Rihan Heights, Abu Dhabi, UAE

PTW Architects
North-Milsons Point, Sydney, Australia

de Architekten Cie.
Tirana Station Towers, Albania

Keith Williams Architects
Black Prince Road, London, UK

Guida Moseley Brown Architects
Helios Residences, Singapore
Draycott Park Residential Complex

arons en gelauff architecten
Golden Gate, Amsterdam, the Netherlands

Ong&Ong Pte Ltd
Alba, Singapore
The Verv @ RV

NL Architects
Prisma, Groningen, the Netherlands
De Kameleon, Amsterdam, the Netherlands

Oppenheim Architecture + Design
3 Midtown, Miami, USA

Special thanks to the architects above for their high-quality projects and nice supports, Inquiries or suggestions are welcome at any time to:
news@hkaspress.com

特别鸣谢以上设计公司的一贯支持，为本书提供优质的作品，如有任何问题或建议请联系：
news@hkaspress.com

图书在版编目（ＣＩＰ）数据

突破风格与复制 II . 上 / 香港建筑科学出版社编 .
—天津：天津大学出版社，2013.6
　ISBN 978-7-5618-4705-3

Ⅰ . ①突... Ⅱ . ①广... Ⅲ . ①住宅 - 建筑设计 - 世界
Ⅳ . ① TU241

中国版本图书馆 CIP 数据核字 (2013) 第 119850 号

责任编辑　郝永丽
装帧设计　黄　丹　王双玲　伍镜光
文字编辑　罗小敏　纪文明　黄夏炎　黄乐琪
流程指导　陈小丽
策划指导　高雪梅

突破风格与复制 II（上）

出版发行　天津大学出版社
出 版 人　杨欢
地　　址　天津市卫津路 92 号天津大学内（邮编：300072）
电　　话　发行部 022-27403647
网　　址　publish.tju.edu.cn
印　　刷　利丰雅高印刷（深圳）有限公司
经　　销　全国各地新华书店
开　　本　245 mm×325 mm 1/16
印　　张　46
字　　数　657 千字
版　　次　2013 年 9 月第 1 版
印　　次　2013 年 9 月第 1 次
定　　价　736.00 元

凡购本书，如有质量问题，请向我社发行部门联系调换